Discovering the Natural Laws

Discovering the Natural Laws
The Experimental Basis of Physics

by Milton A. Rothman

Illustrations by the Author

DOVER PUBLICATIONS, INC.,
New York

Chapter 2 of this book is an expanded version of an article that first appeared in *The Physics Teacher,* Copyright © 1970 American Association of Physics Teachers.

Copyright © 1972 by Doubleday & Company, Inc.
New material copyright © 1989 by Milton A. Rothman.
All rights reserved under Pan American and International Copyright Conventions.

Published in Canada by General Publishing Company, Ltd., 30 Lesmill Road, Don Mills, Toronto, Ontario.
Published in the United Kingdom by Constable and Company, Ltd., 10 Orange Street, London WC2H 7EG.

This Dover edition, first published in 1989, is an unabridged, enlarged republication of the work first published by Doubleday & Company, Inc., Garden City, New York, 1972, in its "Science Study Series." For the Dover edition, the author has written a new chapter (Chapter 10), added references to that chapter, and prepared a separate index to that chapter.

Manufactured in the United States of America
Dover Publications, Inc., 31 East 2nd Street, Mineola, N.Y. 11501

Library of Congress Cataloging-in-Publication Data

Rothman, Milton A.
 Discovering the natural laws : the experimental basis of physics / by Milton A. Rothman ; illustrations by the author.
     p.   cm.
 Reprint, with new ch. 10. Originally published: 1st ed. Garden City, N.Y. : Doubleday, 1972.
 Includes bibliographical references.
 ISBN 0-486-26178-6
 1. Physical measurements. 2. Physics—Experiments. I. Title.
QC39.R68   1989
530—dc20   89-17055
                                                                         CIP

ACKNOWLEDGMENTS

Chapter 2, in a more condensed version, has previously appeared in *The Physics Teacher*. I would like to thank Dr. Clifford Swartz, Editor of that journal, for permission to use that material.

I would also like to give thanks to the libraries of Trenton State College and Princeton University for their hospitality, to Miss Nancy Bemarkt for much manuscript typing, to Mr. Gerald Nicholls for many discussions, and above all to my family for their patience and forbearance, a necessary ingredient of every book.

Contents

Introduction		ix
1	How Do We Know What We Know?	1
2	Definitions, Experiments, and the Laws of Motion	13
3	The Four Forces of the Physicist	48
4	The Gravitational Force	62
5	The Conservation Laws: Historical Background	83
6	The Conservation Laws: Modern Experiments	101
7	The Principle of Relativity	131
8	Electromagnetism	160
9	What Is Forbidden?	183
10	Epilogue (1989)	201
Appendix I Wavelengths, Frequencies, and Energy		215
Appendix II Energy and Momentum Relationships for Moving Particles		218
Appendix III The Michelson-Morley and Kennedy-Thorndike Experiments		223
References		231
Index		237
Index to Chapter 10		242

Introduction

Every book should have a justification. Why should I add another volume to the hundreds of works already devoted to the fundamental laws of physics? The reason became clear to me when—after a considerable amount of writing and teaching about the laws of nature—I suddenly realized that I really did not know how accurately the law of conservation of energy had been verified, and I did not know what were the latest and most accurate experiments testing the validity of this law. This lack of knowledge on my part led me to suspect that many other professional physicists were in a similar state of ignorance. If that is the case, pity the poor science teacher who tries to justify to his students a belief in the absolute truth of the great laws of nature. Either he must present the laws in an authoritarian fashion as being handed down from on high, or else he searches through the textbooks to find some historical experiments by Galileo or Newton or Joule. From these imperfect experiments he must extrapolate to a perfect belief in a perfect law.

The confusion has been compounded by the long controversy concerning the mental processes that take place during the act of inductive inference. That is, how does a scientist arrive at a law such as conservation of energy when the experiments on which the law is based are of less than perfect accuracy and precision? The question is supposedly answered by discussions of how we reach general conclusions after making a sufficient number of specific observations. The im-

Discovering the Natural Laws

plication made by many scientists and philosophers is that somehow we *do* have knowledge that energy is perfectly conserved. But if we look with a realistic and critical eye at what physicists actually do, we suddenly realize that as a matter of fact we *do not* have the knowledge that energy is perfectly conserved. We may think we do, but that is another story entirely. And if, in fact, the law is perfectly true, we have no way of knowing it.

In this matter of understanding the logic of validation, the psychologists have been ahead of the physicists. When a psychologist performs an experiment, he knows that his results are going to be imperfect, and he will couch his conclusions in a manner somewhat like this: "Conservation of energy is a hypothesis that has been verified at a 0.001 level of significance." Which means that there is a 0.1 percent chance that the hypothesis is false.

There have been times when some theoretical physicists tried to lead us into perfect belief by arguments based on the concept of symmetry. We know that each conservation law is associated with a certain symmetry of space and time. For example, momentum is conserved if the properties of space do not change as you move from one place to another. Angular momentum is conserved if the properties of space do not change as you rotate from one position to another. The parity of a system is a property that has to do with the symmetry of the system on both sides of a center line. When a system obeys the law of conservation of parity, we mean that if this system is reflected in a mirror it is impossible to distinguish the mirror image from the "real" system. More technically, if the law of conservation of parity holds true in a given physical reaction, then the description of that reaction is completely unchanged if the directions "right" and "left" are interchanged in that description.[1,2]

The type of argument that used to be stylish during the thirties claimed that since nature liked symmetry (as well as simplicity and elegance) it followed that these conservation laws had to be absolutely true. Such arguments met

Introduction

their downfall in 1957 when the Nobel Prize winners T. D. Lee and C. N. Yang cast a cold eye on the experimental evidence and pointed out that in certain types of elementary-particle interactions involving the weak nuclear force, parity was *not* conserved. This meant that conservation of parity was not a universal law, and that it was possible to distinguish between left and right in an absolute way.[3]

Clearly, it is not possible for nature to like or dislike anything. It is *people* who like symmetry, and we obtain our knowledge of symmetry in nature from observations that show certain quantities (momentum, energy, etc.) to be conserved. We do not obtain knowledge of the conservation laws from an *a priori* belief in symmetry. It is very important to keep this distinction in mind.

Since experimental evidence is the foundation of our knowledge of the laws of nature, it appears that there is a need for a compilation of the best evidence supporting these laws. This book is an attempt to describe a number of the experiments dealing with the most important of the laws. It is by no means a complete and exhaustive work; such a project would be encyclopedic.

Although the experimental data are the raw materials for our theories, such data cannot be understood without a certain amount of earlier theory. A complex feedback spiral involving both theory and experiment is necessary in order to build up a structure of knowledge, approaching but never reaching completeness. For this reason I include some discussion of the philosophy behind the verification of scientific laws. In this area I acknowledge the strong influence of Karl R. Popper, whose ideas have greatly clarified for me the psychology of scientific discovery.[4]

In the final chapters of this book I have attempted to show how knowledge of the fundamental laws may be applied to a critical examination of certain prevalent beliefs. You might find it odd that I include topics such as ESP and astrology, but it is my opinion that there is no dividing line between science and everyday life—if the laws of nature work in the

Discovering the Natural Laws

laboratory, they must work everywhere else. Unfortunately, it is hard for a person to argue against many common superstitions unless he has in his mind a very clear knowledge as to the origins of his knowledge. If you are going to argue with somebody, it does no good to say that "conservation of momentum and energy proves that your idea is wrong." You must know the evidence for your belief in the conservation laws. You must know how good the evidence is, under what conditions the experiments were done, and how accurate they were.

Generally only specialists possess such information. However, I believe that it is important for each person to have some knowledge as to the origin of his beliefs—for if you have a belief in your mind without knowing how it got there, then you run the danger of sliding into unsupported theories, fantasies, superstitions, and finally into the extreme of paranoid delusion. Science is a structure of nondelusional knowledge, and this book tries to point out a few of the foundation stones.

The readers of this book are expected to have a certain familiarity with physics. It may be read with profit by a physics student seeking certain details not given in the usual textbook, and in particular I hope that this book will be found useful by teachers who would like to present to their students the concrete evidence on which our fundamental laws of science are based.

CHAPTER 1

How Do We Know What We Know?

1. *Guess and Test*

How do you know that energy is never created or destroyed? All the textbooks tell us that this is so, glorifying the principle by naming it Conservation of Energy. Furthermore, the books never hint of the possibility of error—the law is obeyed 100 percent. How can we be sure of this? In a universe filled with billions of stars, with uncounted nuclear reactions taking place constantly, transforming trillions of kilowatt-hours of energy from one form into another—how can we be certain that here and there an infinitesimal amount of energy is not lost from the cosmos? Or that somewhere a tiny bit of energy is not created? Who would know the difference?

Scientific knowledge is supposed to be based on observation—on experimental evidence. What, precisely, is the evidence on which our belief in conservation of energy is based? Furthermore, how accurate is this evidence? We know that there is always a limit to the precision of our measurements. How is it that we can speak of perfect laws of nature with absolute certainty when the evidence upon which they are based is not completely certain?

These are questions that cannot be answered entirely within the framework of science, but require us to leap into the more rarefied atmosphere of philosophy. Within the world of science we can deal only with procedural questions: What kind of experiments were done to arrive at the law of conservation of energy? What were the limits of error

Discovering the Natural Laws

on these experiments? What are the most accurate experiments ever performed?

From consideration of experiments with a cargo of human and mechanical error, we then make a logical jump, arriving at a principle of nature that we believe to be perfectly valid —with no error at all. When we ask what right we have to do this, we enter the area of philosophy known as *epistemology* —the subject that deals with the question, *How do we know what we know?* This question has to do not only with the results of complex scientific experiments, but also deals with the most simple and ordinary experiences.

To understand what a complex process lies behind the simplest act of seeing and feeling, think of this book that you are holding. You believe that because you can feel its weight and texture, and see its form and color, you are getting a direct perception of this book. Yet remember that the part of you actually aware of the book is buried somewhere inside your brain. In no way is this awareness mechanism ever in direct contact with the outside world. No pictures are displayed on a screen somewhere inside your head. Connection to the outside world is through the sense organs—the eyes, the ears, and assorted nerve endings. Signals come to your brain by way of those incredible channels of communication, the nerves. Cut the optic nerve and darkness descends.

But what is it that passes through the nerves to the picture-making, concept-forming awareness mechanism? It is nothing but pulses—coded signals transmitting information through a system that baffles our attempts at understanding. The information entering my brain, manipulated to produce the image and feel of the typewriter under my fingers and the sound of my inner voice speaking silently to myself— these bits of information consist of electrochemical pulses and pieces of digitally coded molecules that somehow tell me what is going on "outside."

The essential core of the mystery is the isolation between the conscious part of the brain and the outer world. To understand how this consciousness operates is the ultimate

How Do We Know What We Know?

problem of science, and at this moment we have barely begun the task of solving the problem. In the meantime we casually assume that consciousness does somehow work, that the pulses parading down the nerves become somehow translated into a more or less realistic representation of what is "out there."

But the moment we say this we realize what a big catch there is in the "more or less" qualification. If I hold a simple red sphere in my hands, immediately an interpretation of form is required. Looked at with one eye, the sphere cannot be distinguished from a disc. With only that information to go by, my awareness mechanism must make a guess: This thing in my hand is either a disc or a sphere, and I guess that it is a sphere. With the additional information added by stereo vision and the touch of my hands, I am able to test that guess and thus build up in my mind the geometrical concept of "sphere."

Others making the same tests agree with me that the object is a sphere, and we all agree that the color is red, although of course there is no way for me to know whether your sensation of red is the same as my sensation of red—we can only agree on the distribution of wavelengths in the reflected light.

Even with the simplest observations, then, the little creature inside my head is required to make guesses about the nature of things outside, and common everyday thought becomes a continuous series of guessing and testing. Even when I look at a familiar orange and know from past testing and feeling that the orange has a spheroidal shape, I am making the guess that this orange is essentially the same as past oranges.

Millions of years of evolution have brought us to the point where with practice we can make reasonably accurate judgments concerning the external form of objects close enough to touch. But as soon as we look at a star or satellite beyond reach, as soon as we try to understand the structure of a crystal, an atom, a living cell, as soon as we try to perceive the laws that govern the motion of stars and of elementary

Discovering the Natural Laws

particles, then we are forced to start making more guesses. If the guesser is a scientist and he has the intention of testing these guesses by a suitable set of observations, then he dignifies each guess by calling it a "hypothesis."

Every person engaged in experimental research knows that much of his creative effort appears in the form just described. Imagine a physicist doing an experiment in which he produces a plasma—a volume of ionized gas—by sending a huge blast of electric current through a container of hydrogen gas located within the coils of an electromagnet. He sees a flash of light that he analyzes with a spectroscope, he obtains electrical signals from tiny probes immersed directly in the plasma, he obtains other signals from coils of wire wrapped around the chamber. All of these signals are displayed on the screen of a cathode ray oscilloscope, and as he watches the wiggles, the oscillations, the complex gyrations of the oscilloscope trace, the experimenter must interpret these signals to form a picture of the processes and events taking place inside the discharge tube. In his mind's eye, he uses his knowledge of the motion of electrons and protons in electric and magnetic fields and tries to build up an image of the complicated waves and turbulences taking place within the heated ionized gas.

Without a theoretical idea to start with, the raw signals are meaningless. As philosopher Karl Popper says: ". . . our ordinary language is full of theories; . . . observation is always *observation in the light of theories* . . ."[1] In any kind of experiment where the investigator is trying to find out what is happening inside a "black box," his mind progresses through a continuous series of guesses interleaved with tests using data either from the present experiment or from previous ones. This "guess and test" method is one of the important features of scientific method in general. Conscious or not, the method is used not only to find out what is happening during a particular experiment, but most importantly, to arrive at the great universal laws of nature.

How Do We Know What We Know?

2. *Newton and the Law of Gravitation*

As an example of how the guess and test method works, we might try to re-create a few of the thoughts that went through Newton's mind while he was developing the law of gravitation. He had already decided that objects tend to travel in a straight line if left to themselves, and in his second law of motion he had made the hypothesis that a force is required to change the motion of an object. (See the next chapter, as well as Newton's *Principia* for a more detailed discussion of the laws of motion. The discussion in this section is grossly oversimplified, as a study of the *Principia* will reveal.) Having established the concept that an acceleration requires a force of some kind, Newton was now faced with the embarrassing observation that the moon clearly went around the earth in a curved path; it was continuously undergoing acceleration toward the center of the earth, and yet there was no apparent force acting on it.

Here is where the apple enters the picture. Seeing the apple fall from a tree on his Woolsethorpe farm, Newton's attention was attracted by this event, setting his mind into high gear, causing him to associate two thoughts that had previously been unrelated. It had long been known that "gravitation" was the force that caused apples and other objects to fall to the earth. What Newton sought was a force that could cause the moon to fall toward the earth. Leaping the gap between the familiar and the unfamiliar, Newton made a bold guess: The force that pulls the moon toward the earth is exactly the same as the force that pulls the apple out of the tree. This conjecture did not tell him what the ultimate nature of the force was. It did not enable him to explain an unknown by means of a "known"—for he still did not know the cause of gravitation. It merely allowed him to use a familiar concept in a new context.

Having made the hypothesis that the gravitational force is the cause of the curvature of the moon's orbit, he then went farther and assumed that the very same kind of force was

Discovering the Natural Laws

responsible for the planets curving in their orbits around the sun. We see how one guess piles on top of another guess. Now it is time for testing. Testing means that you predict a specific result arising from your hypothesis, and you compare this specific result with observations and measurements made on real physical objects. If the measurements agree with your hypothesis, then your theory is a good one—at least for the time being. If the measurements disagree with your predictions, then your theory is not a good one and you must make a new theory, or at least modify the old one until it works.

Fortunately for Newton, there were already at hand measurements that had a bearing on the law of gravitation. These were Tycho Brahe's observations of the planetary orbits. In addition, Johannes Kepler had already searched the data and had discovered a most interesting numerical relationship between the radius of each planet's orbit and the time required for the planet to go around once in its orbit. This relationship is as follows: If we let R stand for the radius of a planet's orbit, while T is the length of its year, then for all planets the ratio R^3/T^2 is the same number. This relationship is the well-known Kepler's Third Law of Planetary Motion.

While this relationship is commonly assumed as being derived strictly from observational data, with no theory involved, even here we find guesswork at play. Certainly Tycho Brahe could not measure R and T with perfect accuracy, and yet in his formula Kepler guessed that the exponent of R should be *exactly* 3, and the exponent of T should be *exactly* 2. The psychological reason for this guess was that Kepler *wanted* to find a simple numerical relationship governing the motions of the planets, and as soon as he found something that came close to being a simple relationship, he jumped to the conclusion that the relationship was exactly true.

Newton, then, proceeded to calculate just how the strength of the gravitational force on the planets must vary with distance from the sun in order for the orbits to agree with Kep-

How Do We Know What We Know?

ler's guessed-at relationship. He found that if Kepler's law was true, then the gravitational force must vary inversely with the square of the distance from the sun. Having established this rule for the planets, the next step was to show that this very same rule applied to the force between the earth and the moon. Here was a prediction that could be tested; to accomplish this, Newton (in his own words) ". . . compared the force requisite to keep the Moon in her Orb with the force of gravity at the surface of the earth, and found them answer pretty nearly."[2]

Notice that Newton realized the data fitted the theory "pretty nearly." Not perfectly. Yet he guessed that in this inverse-square law of gravitation, the exponent of the radius R ought to be exactly 2, not 1.99 or 2.01. It seemed like a reasonable guess, and it satisfied the psychological need for simplicity and form. Furthermore, Newton was able to make a remarkable prediction as a result of this guess. By the use of his newly-invented mathematical method—the calculus —he was able to show that if the strength of the gravitational force follows an inverse-square law, then the orbits of the planets must be exactly elliptical. (The derivation of this theorem is given in any book on mechanics.) Furthermore the orbits are elliptical only if the exponent of the radius R in the law of gravitation is exactly 2—any other exponent gives a different kind of curve. Here was a prediction that could be tested by observation, and indeed these observations had already been made by Tycho Brahe and formulated in Kepler's first law of planetary motion. The orbits of all the planets were actually ellipses, as closely as could be measured.

As measurements of the planetary orbits became more and more accurate over the years, it was found that Newton's guess was indeed a good one. You might say that's just as it should be, but as it turned out, when the measurements became accurate enough, the inverse-square law of gravitation was not really 100 percent correct; it was only the first approximation to the more exact law, and it remained for Albert Einstein to find a new, more exact formula for the strength of the gravitational force (see Chapter 4).

The point of the story is that Newton's law of gravitation involved a certain amount of guesswork. These guesses were tested by comparing predictions of the theory with observed planetary orbits. The tests were never completely accurate, and so when we ask how could Newton be sure that his law was perfectly valid on the basis of imperfect measurements, the answer is simple: Newton could *not* be sure. He may have thought he was sure, but that's another situation entirely. His conviction was based on the idea that a law of nature ought to have a simple form rather than a complicated form. We see from this example that Newton's choice of the inverse-square law was not simply the result of observations "objectively" tabulated and graphed and analyzed, but was partly based on a theoretical idea that he had in his head from the beginning.

3. *Science as a Conceptual Structure*

Some people look upon science as a structure of theories that allow us to explain how things behave. Other people regard scientific knowledge as an orderly arrangement of experimental observations from which we build up general rules that we call Laws of Nature. Our point of view in this book is that science is a very complex structure of mental concepts, hypotheses, theories, and observations, all interlocking in such a way that it is often very difficult to separate theory from observation. As we have seen, even the simplest observation requires an interpretation based on some sort of theoretical idea. Every beginning student can see that the sun circles about the earth and can feel that the earth is standing still. But to recognize that the sun is the center of the solar system requires indoctrination into a more sophisticated mode of thinking—a structure of concepts that separates modern man from medieval man.

It is for this reason that we spend so much time in this book discussing philosophical theory, even though our main inten-

How Do We Know What We Know?

tion is to describe the experiments behind the laws of nature. Before you can understand what a particular experiment proves, you must have a clear idea in your mind concerning the theoretical ideas underlying the experiment. As we will see in the next chapter, even as basic a topic as Newton's three laws of motion is fraught with ambiguities. Such a simple experiment as pulling a car by means of a spring can have more than one interpretation, depending on the conceptual structure built up in the mind of the interpreter.

Often the structure of science is described as being divided into two parts: inductive and deductive. The inductive part has to do with building up generalized laws of nature from a series of observations and experiments. For example, suppose you do a number of experiments in which you measure the amount of energy going into a reaction and the amount of energy coming out; you find that the amount of energy in the system is the same after the reaction as it was before the reaction. According to the inductive point of view, if you do a large enough number of experiments like this, you can jump to the conclusion that in all experiments the amount of energy within a closed system is a constant.

Once you have obtained this generalized law by induction, you can use this law to predict the results of experiments that might be performed, or the operation of devices that might be built in the future. Obtaining particular examples from the general law is the process of *deduction*. Thus, once we believe the law of conservation of energy, we can look at some particular machine and we can say that if it is supposed to give out more energy than was put into it, its operation is impossible.

Karl Popper has criticized the logic of this procedure, offering an alternative way of describing how scientific discovery takes place. Popper recognized that every experiment starts with a hypothesis. After all, the very act of doing an experiment implies that the experimenter has made a choice between performing one operation rather than another. In or-

Discovering the Natural Laws

der to have a hypothesis there must have been some previous —perhaps casual or unorganized—observations. A good hypothesis is one that makes predictions that can be tested. This implies that an experiment can show the hypothesis to be either true or false. It's pointless to make a hypothesis whose predictions make no difference one way or the other. For example, if I make the hypothesis that gravity is caused by an invisible, undetectable fluid called "gravitol," there is no way to test this guess because it doesn't predict any specific properties of gravitation that can be compared with the observed data. Therefore it doesn't make any difference whether I believe this theory or not. To make a good hypothesis I would have to list properties of gravitol that would cause the force of gravitation to behave as actually observed. Even then there is still the possibility that gravity could be explained just as well by some other theory.

When making a hypothesis such as that of conservation of energy, the most important requirement is a definition of energy that allows its quantity to be measured. (For example, Energy = Work = Force × Distance.) Then I can test the hypothesis by measuring the quantity of energy within a system while various reactions are going on. If I find that this quantity of energy remains constant every time I make this measurement, I shall feel that this hypothesis is a good one. Perhaps the experiment was done only ten times. But if I have faith and imagination, I make a leap of fantasy in my mind—I say let's suppose that in *any* kind of system the total amount of energy is a constant. This means that whenever any kind of machine is built it will be found that the amount of energy coming out of this machine is exactly equal to the energy going into it.

The years go by, different kinds of machines are made, more and more accurate measurements are made. Soon a difficulty arises: I put a lot of mechanical energy into a machine, and only half of this comes out, together with a lot of heat. Do I say that energy is lost? Do I say that my hypothesis is falsified? No. I find that I can save the hypothesis by de-

How Do We Know What We Know?

fining the heat itself to be a form of energy. This can be done because it is found that a given quantity of heat always corresponds to the same quantity of mechanical energy (work). Thus, by broadening the definition of energy to include heat, I fix things so that my hypothesis is not falsified; energy is once more conserved. As time goes on, new experiments are done, new machines are built, and there are many occasions where the definition of energy must be broadened to include new forms—electromagnetic, nuclear. And at one point it is even necessary to invent an invisible, almost indetectable particle, the neutrino, in order to retain the hypothesis of energy conservation. Each new definition, each new form of energy and matter becomes another girder in the conceptual structure built in my mind.

Popper calls this method of creating a law of nature "the deductive method of testing," for it involves starting with a few observations, making a hypothesis to explain the observations, deducing further observable events from the hypothesis, and then testing the hypothesis by making more observations to see if they agree with the predicted events. This procedure is nothing more than the "guess and test" method described in the preceding section. An important feature of this method is the idea that no number of positive verifications can prove the hypothesis to be absolutely true, but it only takes one negative experiment, or falsification, to demolish the hypothesis.

After all, no matter how many millions of experiments we do to verify a law such as conservation of energy, we cannot predict what might happen under strange and unusual conditions such as those in the center of a large star, in a very distant galaxy, under extremely high temperatures. We are only sure what happens under those conditions where we have done experiments to measure energy transformations. Even then, all we can say is that within certain limits of experimental error the amount of energy does not change.

Furthermore, an even more fundamental assumption is involved, and that is: Whenever a measurement is made under

Discovering the Natural Laws

a certain set of conditions, then the same kind of measurement under the same conditions will give the same results. In other words, once conservation of energy has been tested with one kind of reaction, then it will always be a good law for that kind of reaction. A law does not change with time.

On the other hand, if you ever find one kind of situation where energy is definitely lost or created, then the law is overthrown—at least for that particular kind of situation. So far this has not happened, but we cannot take it for granted that it will never happen.

All of the difficulties involved in verifying a hypothesis apply, of course, only to guesses about very general predictions. If you make a guess about the number of beans in a pot, it's very simple to count the number of beans to verify your guess. But if your guess happens to be that every bean pot in the world holds less than ten thousand beans, then no matter how many pots you count, you will not be sure what the next pot might have in store for you, unless you investigate every pot in the world. The first pot you find with more than ten thousand beans will prove the hypothesis wrong.

Fortunately, we don't have to observe the properties of every electron in the universe to be reasonably sure of the laws that govern the behavior of electrons—as well as the other elementary particles. All electrons are alike, and once we have observed enough electrons to know how they behave, then our guess is that all the electrons in the universe behave the same way. If we meet a particle that is somewhat like an electron but behaves differently we are inclined to call it another kind of particle and give it a different name. The fact that all particles of a given type behave identically is fundamentally what allows us to make reasonable hypotheses that become the framework of our conceptual structure about the nature of the universe.

CHAPTER 2

Definitions, Experiments, and the Laws of Motion

1. Definitions of Force and Mass

We—and especially Sir Isaac Newton—are fortunate that kilogram masses do not behave like nuclear particles such as neutrons and protons. Imagine this situation: Newton is experimenting with a linear air-track—a device (commonly used in laboratory teaching) that allows a car to slide along a compressed air cushion with little friction. He ties a string to one of these cars, and hangs the string over a pulley at one end of the track, so that by suspending a small weight from the end of this string he obtains a constant force applied to the car (Figure 2-1). Loading a one-kilogram mass on the car,

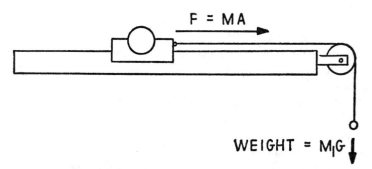

Figure 2-1. A linear air track, with a car gliding frictionlessly over an air cushion maintained on an aluminum beam. The way the car's acceleration depends on the mass of the car and the magnitude of the applied force can be observed with minimum error due to friction.

Discovering the Natural Laws

he proceeds to measure the car's acceleration and finds, let's say, one unit of acceleration.

Now he adds a second one-kilogram mass to the first already on the car and repeats the experiment. He has doubled the number of masses being accelerated, and so he expects to find half the acceleration. But to his astonishment he finds, instead of 0.5000 unit of acceleration, a greater value: 0.5005 unit.

If he tries to write the equation

$$\text{Force} = \text{Mass} \times \text{Acceleration}$$

for each of the individual masses, he cannot write the same equation for the two masses combined unless he admits that two masses do not add together in a simple, linear way. One kg plus one kg does not equal two, but instead equals 1.998 kg. With matter behaving like this, the laws of motion become rather complicated and their discovery by Newton becomes much more difficult.

An unlikely story, you say. On the contrary, this kind of complication is just what the nuclear physicist puts up with every day. When he applies a force to a rapidly moving charged particle by means of a magnetic field, he expects the radius of curvature of the particle's path to be directly proportional to the mass of the particle. In this way he can compare the masses of various particles. Now suppose he measures the radius of curvature of a proton's path in a given magnetic field. He then makes the same measurement using a beam of deuterons, each of which is nothing more than a proton joined to a neutron. When he does this, he finds that the bending of the deuteron's path is about a tenth of a percent greater than he would expect by adding the mass of the proton to the neutron mass. Does he conclude from this that Newton's Laws of Motion are not valid? Does he say that the behavior of the magnetic field becomes different when it acts on a proton attached to a neutron? No, he simply says that the proton and neutron locked together have less mass than when they are separate. Furthermore, when

Definitions, Experiments, and the Laws of Motion

he looks closely enough at what is happening, he sees that a gamma ray photon left the scene when the neutron and proton combined, and he explains the situation by saying that this photon carried some mass away from the deuteron as it was being formed.

If kilogram masses behaved in the same manner as the neutrons and protons we have been discussing, Newton would have found it difficult to express his second and third laws of motion in a simple way. But let's think about this carefully. What gives us the right to assume that kilogram masses do *not* behave like neutrons and protons? After all, they do attract each other with a gravitational force (together with any cohesive forces that might exist at their interface), and if the gravitational binding energy is allowed to escape from the system when the two masses are brought together, then there will be a loss of mass. What makes us —and Newton—so lucky is the fact that the gravitational binding energy is so infinitesimally small that there is no possibility of noticing it in any laboratory or astronomical situation. Therefore we can get away with the assumption that two equal masses put together add up to twice the individual masses—that is, they add in a linear manner (the total mass plotted against the number of masses is a straight line).

This story has the moral that no assumption can be taken for granted. The only reason for believing that masses add linearly is the fact that this is the way we see they behave. Otherwise, there is no *a priori* reason for assuming that physical masses add together in the same manner as arithmetic whole numbers. We are not free to define the behavior of masses the way we can define the behavior of whole numbers in pure mathematics.

We learn from this experience that the moment we try to reach more than a casual understanding of Newton's laws of motion we find hidden behind those familiar words a complex structure of hypotheses, assumptions, and experimental observations, much of which is never brought into the light of awareness, even in the most careful textbooks. Furthermore, we find that there is a great amount of confusion as to what parts of the laws are simply definitions and what parts are based on experiments and observations.

This state of affairs has been well known for many years, and an entire literature has grown up around the attempt to put Newton's laws of motion on a logical and self-consistent footing. Much of this literature is summarized in *Resource Letter on Philosophical Foundations of Classical Mechanics*, by Mary Hesse.[1] Among the classic writers dealing with the subject are Ernst Mach,[2] Henri Poincaré,[3] and Hans Reichenbach.[4] The treatises of Leigh Page[5] and Lindsay and Margenau[6] have exceptionally complete and logical analyses of the definitions of mass and force, while Max Jammer[7] devotes a volume to each of these concepts. More recently Leonard Eisenbud[8] and Robert Weinstock[9] have published detailed analyses of the laws of motion because of dissatisfaction with the customary treatments. Eisenbud, in fact, expressed his dissatisfaction by replacing Newton's three laws with a set of his own designed to have a more rigorous structure than the original three.

Among the more traditional, engineering-oriented textbooks the easy way out of the complex situation was to quote the three laws of motion as Great Laws of Nature handed down to mortal man from higher authority. The trend among the better modern books is to present a reasonable amount of the background and logic behind Newton's laws. However, hardly ever does one see written down a step-by-step sequence of definitions, hypotheses, and experimental tests leading up to the laws of motion in the spirit of Euclid's geometry, leaving no holes in the chain of logic.

One reason for this state of affairs is the fact that there is no general agreement on one best logical sequence. Indeed, there are some violent differences of opinion on the best way to approach the subject. It seems strange that in a subject as old and presumably settled as classical Newtonian mechanics there should still exist strenuous debate concerning the fundamental concepts underlying the entire field of study.

Definitions, Experiments, and the Laws of Motion

Yet, as we shall see, there are at least two common and important terms (*force* and *inertial reference frames*) that are given varying definitions by the leading textbooks in use, giving a clue that the subject may not be as settled as you might think.

A fundamental reason for these differences of opinion is the fact that the laws of motion are not absolute laws handed down by higher authority, but rather represent man's attempt to build in his mind a conceptual structure that enables him to interpret what goes on out in the world. As we described in the last chapter, it is the oldest problem in philosophy: How do we know what the universe is like when all that comes into our brains are coded sequences of pulses traveling along a number of nerve channels originating in our sensory organs? We do this by building up in our minds a set of concepts such as acceleration, force, mass, atoms, molecules, and nuclei. The organization of these concepts into a unified structure is the business of science.

Trouble arises when different people invent different structures, all claiming to describe the same things existing "out there." Newton's statement, "$F = ma$" is a simple thing, but there are at least three ways of interpreting it. The traditional engineering approach stems from the muscular feeling you have when you push or pull something. As a result, the engineer begins by defining the unit of force to be the pound force (represented by the weight of a standard pound at a given spot on the earth's surface). The unit of mass is then defined as that mass that responds to the pound force with an acceleration of one foot/sec^2, and we end up with that ungainly unit of mass, the slug.

The school of physicists following the philosophy of Ernst Mach notices that forces are not really the results of direct contact between hard objects, but that they are transmitted through empty space through "fields." As a result the point of view is changed, and we say: Here we see an object undergoing acceleration, so there must be a force acting on it, even though we can't see anything touching it. If we know the

Discovering the Natural Laws

mass of the object, we can define the amount of force by the equation $F = ma$. We are then left with the problem of first defining the object's mass, and Mach solved this problem by saying: We compare an unknown mass with a standard mass by using an inertia balance (Figure 2-2). When we push the

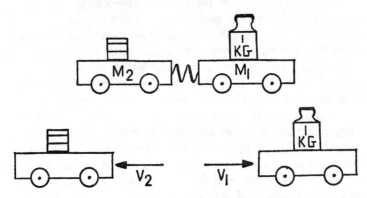

Figure 2-2. An inertia balance. Two cars are pushed apart by a spring (or other force). By comparing the velocities (or accelerations) of the two cars, the inertial masses of the two cars may be compared.

two masses apart by a spring or other force and then measure their velocities (or accelerations), the ratio of the masses is given by the ratio of the velocities, for those ratios represent the relative amounts of inertia of the two objects.

A third approach, that used by Arons,[10] is a compromise between the two. A standard kilogram is defined, and then a spring is calibrated by measuring the acceleration it gives to that standard mass while the elongation of the spring is measured. We then use that calibrated spring as a known force to find the value of unknown masses by measuring their acceleration (Figure 2-3).

In this third approach (which we will call the Newtonian method), as well as in the Machian approach, we compare unknown masses with a standard mass by using a spring. In the Machian method we use the spring to accelerate both

Definitions, Experiments, and the Laws of Motion

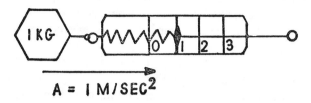

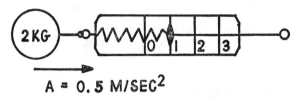

Figure 2-3. A 1-kilogram standard mass is accelerated at a rate of 1 m/sec² with the help of a spring balance. By definition, the amount of force required is 1 Newton, so we can label the point on the scale corresponding to this amount of spring stretch 1 Nt. If we now apply this known force to an unknown mass, we can find the unknown mass by measuring the acceleration and using $m = F/a$, according to the Newtonian point of view.

masses **simul**taneously, while in the Newtonian method we pull the objects one at a time, trusting that the properties of the spring do not change between pulls. However, the psychology between the two methods is quite different. The Newtonian method says, in effect: I have a *known force* that I apply to a mass. If I measure the acceleration this tells me the amount of mass. If I know the mass, then I can predict what the acceleration is going to be. This is indeed the way we go about solving practical problems, when the source of the force is a known quantity.

The Machian method, on the other hand, has a very interesting connotation. It says: If I know the mass of an object and I measure its acceleration, then this operation tells me how much force is causing the acceleration. The force must be given by $F = ma$, by definition, regardless of the source.

Discovering the Natural Laws

This is the method used to discover new forces, and in fact was the method used by Newton when he realized that the acceleration of the moon toward the earth was caused by the gravitational force.

Now, however, we run into a serious logical difficulty. If $F = ma$ is true by definition, then what is the point of all the experiments we perform in the classroom to demonstrate that the laws of motion are "true"? When we try to answer this question by a close analysis of Newton's laws, we find ourselves going through a thorny philosophical thicket, but when we come out on the other side we find that we begin to understand what a "law of nature" really is, and what experiments must be performed to verify each law.

2. The Laws of Motion

To illustrate the process of analyzing a law of nature, let us try to re-create Newton's laws of motion step by step. I will use the Machian scheme here, although the same procedure can be applied to the Newtonian method. The guiding principle is that no word may be used before it has been defined in some way. In this manner we avoid circular reasoning, the trap that harasses so many discussions of the second law.

One important feature of the analysis used here is that in order to give full meaning to certain concepts I find it useful to employ three different types of definitions: conceptual, behavioral, and operational. A conceptual definition is a description of the concept I have in mind when I use a word, while a behavioral definition tells how I want something in nature to behave or what I want it to do. An operational definition describes the operation to be performed in order to measure the quantity being defined.

Operational definitions are favored by physicists because there are many concepts such as a "unit of length" or "an interval of time" or "energy" that simply cannot be defined properly by words alone, but which can be defined by describing certain operations that satisfactorily measure the

Definitions, Experiments, and the Laws of Motion

thing being discussed. However, sometimes a bare operational definition is a bit empty unless we can explain why we *want* to measure a quantity in a given way. A behavioral definition can put meat on the bare bones of an operational definition, supplying the motivation that gives meaning to the operation.

To arrive at Newton's laws of motion we start with the customary mathematical definitions of position, velocity, and acceleration with respect to a given reference frame, and we define a physical straight line to be the path followed by a light ray in free space. We also bring with us the primitive conceptual definition of a force as a push or a pull.

When we try to make a more sophisticated definition of force, we ask ourselves what it is that a force does, and we see that force is invariably associated with acceleration. As Eisenbud points out, when a force is suddenly applied or removed, the position and velocity of the object do not change suddenly (i.e., they are continuous), but it is the acceleration that changes suddenly, discontinuously. (Even in statics, where there is apparently no motion of any of the structural members, there must be a momentary acceleration of the atoms within the member as the force is first applied. In this way the force is transmitted through the beam, rope, or lever, and the reaction forces are built up by the atoms squeezing together the electric fields between them. Infinitely rigid bodies exist only in the minds of mathematicians and classical physicists, not in nature.)

As a result of these observations we are encouraged to make a behavioral definition of force: *A force is anything that can cause the acceleration of an object.* (This definition is similar to Newton's own Definition IV: An impressed force is an action exerted upon a body, in order to change its state, either of rest, or of uniform motion in a right line.[11])

Now just because we have made a definition, this does not ensure that the defined object exists. Hidden behind every definition is the unconscious hypothesis that the thing being defined does, indeed, exist. Let us bring this hypothesis out into the open and say: If an object is seen to undergo

Discovering the Natural Laws

acceleration, then there exists a force that is acting on it. In other words, an object changes its state of motion if and only if there is a force acting on it.

Unfortunately, when we ask how we might test this hypothesis, we realize that our definition of force is so general that it is untestable, for there is no way to get a negative result (or falsification) out of any physical test we might apply. That is, if we claim that every accelerating object has a force applied to it *by definition*, then there is no way of showing this hypothesis to be false. In the same way we could define into existence angels, invisible crystal spheres guiding the planets around the earth, heavenly chariots, and so on. This type of definition was typical of medieval thinking: define into existence some invisible entity that accounts for the observed phenomenon. Since nobody detects this thing, it doesn't make any difference whether you believe in it or not.

The entry into modern thinking starts when we realize that one more ingredient is necessary for a proper definition of force. The acceleration, and thus the force, must be the result of an interaction of one object with another object, so that whenever we see something undergoing acceleration we can point to something else and say: This is the source of the force acting on the accelerating object.

Including this new idea, we write another definition of force: *A force is something that arises from the interaction between two or more objects and results in the acceleration of each of these objects.* Accompanying this definition is the hypothesis that the acceleration of an object is always caused by such a force.

Now we have an hypothesis that can be tested. If, every time we see an object being accelerated, we can identify the origin of the force responsible for that acceleration, then the hypothesis is a good one. When we make this test we find that there are apparently two kinds of accelerations in the universe. First there are accelerations produced by identifiable sources, such as masses, electric charges, stretched springs, and so on. Upon close examination we find that there

Definitions, Experiments, and the Laws of Motion

are four fundamental types of sources, and so we label the four fundamental interactions, or forces: gravitational, electromagnetic, and two kinds of nuclear force—the strong and the weak.

Second, we find many strange situations where objects are seen to undergo acceleration, or where the observer feels a force acting upon himself, but where there is apparently no source visible capable of producing this acceleration or this force. Yet, as the occupant of a centrifuge can testify, the effects of such a force can sometimes be quite violent.

Is there a general kind of rule that tells us when we might see these "sourceless forces" in operation? In fact there is. We find that such forces are felt only when the observer happens to be in a reference frame that itself is being accelerated—either his frame is rotating so that there is an acceleration toward the center of curvature, or he is in a vehicle that is speeding up or slowing down. These sourceless forces are given the general name of *inertial forces,* and the particular names of centrifugal force, Coriolis force, and the "g-force" felt by the rocketing astronaut (Figure 2-4).

There are three possible things we can do with the information we have obtained while testing our hypothesis about the nature of force. First, we can say that this hypothesis is all wrong, since not all forces have identifiable sources. Alternatively, we can say that our force hypothesis is correct, but is correct only in certain special frames of reference in which inertial forces do not make themselves felt. In these "inertial reference frames" we can always identify any local source of force causing an object to accelerate. In other—noninertial—reference frames, a free object appears to accelerate without cause.

Since our minds insist on associating an acceleration with a force, and since we certainly feel a force of some kind when we whirl around in a centrifuge, we use the term "inertial force" to describe this feeling, but then turn around and deny its reality by calling it a "fictitious force." We then explain to the subject being squashed in a centrifuge that this ficti-

Figure 2-4. An astronaut in an accelerating rocket far from any planet or star feels an inertial "g" force pulling him "downward." A spring balance in his hand registers the "weight" of the steel ball hanging from the balance. This "weight" is the result of inertia, but cannot be distinguished from an ordinary gravitational force. This equivalence of an accelerating reference frame and a gravitational force is Einstein's "Principle of Equivalence."

Definitions, Experiments, and the Laws of Motion

tious force he feels is simply the result of his "inertia," and carefully sweep under the rug the question of why he has "inertia" in the first place.

The third option is to keep open the possibility that the inertial forces actually can be identified as real interactions between the accelerating object and objects elsewhere in the universe. Early in the 18th century Bishop Berkeley expressed the idea that all motion—accelerated as well as uniform—must be measured relative to the universe of fixed stars. In the 19th century Ernst Mach considered the possibility that the inertial forces are due to some kind of physical interaction between the objects undergoing acceleration and the rest of the universe, an idea now known as *Mach's Principle*. Following along this same path we arrive in the 20th century at Einstein's General Theory of Relativity, in which the inertial forces may arise from certain components of the gravitational interaction.

If we want to remain within the framework of classical Newtonian theory we use the second option above, but keep in back of our minds the fact that this is a man-made choice, and that the third option may lead us closer to understanding the way nature actually behaves. Mach's Principle has not been widely accepted, except among some specialists in general relativity. However, if new developments in theory[12] survive the test of observation it will be necessary to rewrite all the elementary textbooks to remove the expression "fictitious forces" from their pages.

Having made at least a partial definition of force, we continue in our scheme to find a definition of mass. In all that follows we will assume that our experiments are performed in an inertial reference frame. The problem of defining an inertial frame (or of determining when you are in one) is another of those difficult points on which there is considerable disagreement. A survey of the textbooks in common use shows that there are two main definitions in favor. One definition states that an inertial reference frame is a frame that is at rest (or moving with constant velocity) with respect to the average position of the distant fixed stars. The second definition is the one we have used in the previous paragraph: An inertial reference frame is one in which no "inertial forces" are felt. That is, an object left to its own devices with no apparent force sources in the vicinity will be seen to move with constant velocity in a straight line if the observer is in an inertial reference frame. (In other words, an inertial frame is one in which objects obey Newton's First Law of Motion.)

The first definition is a pure mental construct, one virtually impossible to put into actual practice by observations on the stars. The second definition is the one actually used to determine whether a given frame of reference is actually an inertial one. (For example, one can construct a sensitive accelerometer for space travel by suspending a metal ball freely inside a closed chamber located somewhere within the space vehicle. If the vessel is traveling freely through space, the ball will not change its position relative to the confining chamber. But if the ship undergoes acceleration due to some external force, the position of the ball will change, and this motion can be detected by instruments located within the chamber.[13])

Notice that the two definitions of an inertial frame are not identical. A space capsule falling freely toward the earth will satisfy the second definition, but not the first. All the things in the vehicle accelerate at the same rate toward the center of the earth, and so they maintain constant positions with respect to each other, if they are at rest with respect to the frame, and they travel in straight lines if they are in motion (as long as you don't make the frame big enough to notice the fact that the gravitational lines of force are not really parallel). The accelerometer described in the previous paragraph will not be able to tell the difference between a ship falling straight down toward a planet or a ship moving with constant speed in a straight line in free space. Both of these frames are, therefore, inertial frames.

For the time being we will adopt the definition that *an inertial reference frame is one in which objects obey New-*

Definitions, Experiments, and the Laws of Motion

ton's First Law of Motion (*whether they are at rest or in motion*). At first blush we appear to have reduced the First Law to a definition of an inertial frame, and we ask what makes it a law of nature. What is the experimental content of the law? The experimental fact is that such frames do exist; we can find reference frames in which an isolated object at rest remains at rest and an object in motion will proceed with constant speed in a straight line. Without this observed fact the definition would be a group of words describing a nonexistent entity.

Now let us put ourselves into such an inertial reference frame and do some experiments. We first observe an object A that is undergoing acceleration, and we notice that this action is associated with the acceleration of another object B in the opposite direction. This observation leads us to say that these accelerations are caused by a force-pair, the interaction between A and B. We may find that object A has a greater acceleration than object B. We interpret this behavior by introducing the concept that B has a greater amount of inertia—resistance to change of motion—than A, leading us to define the inertial mass of an object as a quantitative measure of that property called inertia.

In order to measure inertial mass we must define a standard of mass, which we do in the usual way by pointing to a piece of platinum and saying: This is a standard kilogram mass. We then say: If an object has the same amount of inertia as the standard kilogram mass, then this object has an inertial mass of one kilogram. This is a behavioral definition, which must be completed by an operational definition that tells us how to compare the two masses.

There are many possible operational definitions; I choose one similar to that used by Mach—the inertia balance illustrated in Figure 2-2 (see page 18).

Let us assume that the two masses are the two cars shown in the figure. To make sure that there are no misleading extraneous influences, I have a spring push the two cars apart a number of times, each time with the cars arranged in a

Discovering the Natural Laws

different orientation. Also the experiment should be done in a number of different locations in space. If the two cars end up traveling with the same speed regardless of orientation or position, then by definition the two masses are equal. This procedure establishes a frame of reference with a certain degree of symmetry. Without this precaution we could find ourselves measuring masses that were a function of the direction of motion, an event that might occur if space were not symmetrical in all directions, or—what amounts to the same thing—if there were some unknown source of attraction off in some direction.

This procedure, as a matter of fact, is another method of verifying that we are in an inertial reference frame. I now proceed to the crucial part of the experiment, which is to continue the comparison of the two masses under many different conditions: with different amounts of spring compression, with different kinds of springs, with different kinds of forces, with masses of different materials, and so on. I always find that when two masses have been found equal initially, they continue to be equal when the test is made in all these different ways.

Here we find the point where definitions end and experimental results become important. Initially I defined two masses to be equal according to a certain operation, using one particular spring. I then find experimentally that the equality persists even though I may modify the operation in various ways. Thus we find that the equality of the two masses does not depend on the position, orientation, and velocity of the two objects, and it does not depend on the amount and type of force used to push them apart, as long as the experiment is done in an inertial frame of reference.

Another important property of inertial masses is discovered when we compare three objects with each other. We then find that if $M_1 = M_2$, and $M_2 = M_3$, then $M_1 = M_3$. That is, equality of mass obeys a transitive rule. This discovery is something we could not have established by definition. This is simply the way masses behave.

Definitions, Experiments, and the Laws of Motion

Now, suppose I put two different masses—M_1 and M_2—on the balance (M_1 being the standard kilogram mass) and find that v_2 (the velocity of the second mass) is less than v_1. This means that M_2 has more inertia than M_1. I now define the inertial mass M_2 to be the quantity

$$M_2 = M_1 \left| \frac{v_1}{v_2} \right|$$

This definition satisfies my requirement that the mass be proportional to the amount of inertia, and the operation provides a quantitative way of measuring this amount of inertia. (We could just as well have used the ratio of the accelerations instead of the ratio of the velocities, but that amounts to the same thing.) Notice that this definition uses one particular spring under one set of conditions. Now I find that if I repeat the experiment with all the variations described above (different springs, different forces, etc.) I always find M_2 to be the same. This experimental discovery convinces me that the mass that I have more or less arbitrarily defined by the above formula represents an intrinsic property of the object itself, rather than being an accidental result of one specific operation.

The equation defining the mass can be rearranged so that it reads

$$M_1 v_1 + M_2 v_2 = 0$$

The quantity Mv is defined as the momentum of each object, and we must take care to account for the directions of motion of the two objects. For example, if object 1 goes to the right and object 2 goes to the left, then we make v_2 a negative number while v_1 is positive. The total momentum of the two objects before being pushed apart was zero, and the above equation says that their total momentum after being pushed apart is still zero. We recognize that the momentum of the system is unchanged by the interaction, and since this is true under all conditions we call this relationship a law of nature and name it Conservation of Momentum.

Discovering the Natural Laws

The criticism is often made that if we define mass by the above recipe, then we have made conservation of momentum true by definition because we seem to get the conservation law just by rearranging the formula that we used to define the mass of object 2 in terms of the standard mass. We wonder, then, what is the experimental content of the law? If a law of nature is something we discover by observation of nature, where have observations gone into the construction of this important law we have been discussing?

By looking at the careful analysis of the definition of mass in the above paragraphs, we can see what parts of the law are the result of experiments. We begin by defining the mass of an object under one particular set of conditions. We then find by experiment that the above relationship continues to be true for all time, regardless of the orientation of the masses, their location and motion in space, regardless of the kind of force used to push them apart, and regardless of the materials they are made of. *The invariance of the relationship is the essence of the law.*

Furthermore, the momentum is found to be conserved in all inertial frames of reference. That is, any observer moving with constant velocity will find the total momentum of these two objects to be unchanged by any interaction taking place between them. This is another fact we learn from experiment, and represents the invariance of the relationship regardless of the motion of the observer. (Or, as the mathematicians put it: invariance with respect to a transformation from one inertial reference frame to another.)

Other important relationships concerning the measurement of mass remain to be shown by experiment. For example, if $M_1 = M_2$, then $M_1 + M_2 = 2M_1$ (linearity of addition of mass). This property allows us to set up a simple system of mass units and, as described at the beginning of this chapter, it is a property that cannot be taken for granted or defined into existence.

Having defined mass, we are now ready for the quantitative (operational) definition of force, which we do by say-

Definitions, Experiments, and the Laws of Motion

ing that whenever we see an object being accelerated, the *amount* of force acting on that object is equal numerically to the rate of change of momentum. If the mass is constant, this statement becomes our familiar friend F = Ma—Newton's Second Law of Motion.

Once more the question is raised: What right do we have to call this a law of nature? If F = Ma is true by definition, what is the experimental content of the law? The experimental content is considerable, but not always obvious. First of all we observe that if we do these experiments in an inertial frame of reference a force acting on mass M always arises from an interaction with another object—that is, a stretched spring, an electric charge at a given distance, a planet, etc. Second, we observe that if a particular interaction (e.g., a particular spring stretched a certain distance) causes a force of magnitude F = Ma, then that same interaction under the same conditions will always give the same mass the same acceleration. It is this constancy of the interaction that allows us to calibrate a spring to measure force. Third: If a particular interaction acts on one object with a force $F_1 = M_1 a_1$, and if the *same* interaction acts on another object with a force $F_2 = M_2 a_2$, then it is found that $M_1 a_1 = M_2 a_2$, and so $F_1 = F_2$. That is, the same spring produces the same force (for a given stretch) regardless of the object it is accelerating. It is this property that allows us to use calibrated springs to exert known forces on all kinds of objects.

In general, if a given arrangement of objects produces a certain force, this same arrangement will always produce the same amount of force. For example, if we arrange a number of electric charges on a pair of parallel metal plates (e.g., the deflection plates of a TV picture tube) and pass a beam of electrons between these plates, the first time we do this we notice that the path of the electron beam is deflected. This motion shows us that there is a force acting on this electron beam, and we can use the measured deflection to calculate how strong the force is. Thereafter, knowing from experience that the same cause is going to give the same effect, I

Discovering the Natural Laws

can turn the logic around and say: If I make the same arrangement of charges on the deflecting plates in the future, this will cause the beam to be acted on with a known force, and now I can predict how far the beam will be deflected.

A fourth and most important observation is this: If a number of forces act on an object, we will observe that the forces add linearly according to the rules of vector addition—the well-known polygon or parallelogram rules for addition of vectors.

We see that the essence of the law is the invariance of the

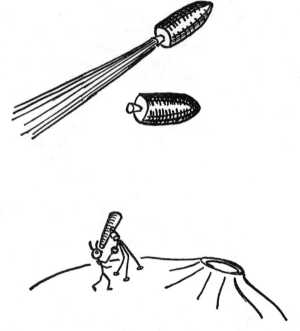

Figure 2-5. A rocket is seen to be accelerating by the observers in another spaceship. The observers on a nearby planet and the observers in the spaceship find the rocket to be moving with the same rate of acceleration, even though these two sets of observers are moving with constant velocity relative to each other. Velocities are relative, but accelerations appear to be "absolute."

Definitions, Experiments, and the Laws of Motion

relationship $F = Ma$. Once a force is defined by this relationship, it remains unchanged. The experimental tests of the law must deal with the invariance properties and the summation properties, not with the original definition of force. Another question that must be answered by experimentation is this: If an object is accelerating under the influence of some force, as seen by one observer, how does the situation appear to another observer moving at constant speed relative to the first observer? (see Figure 2-5).

The answer, according to classical Newtonian physics, is that the object's mass and acceleration are found to be the same by both observers, and therefore the force is independent of the motion of the observer relative to this object. Similarly, the total momentum of a system of interacting objects is found to be the same for all observers. These observations deal with the transformation properties of the laws. The usefulness of Newton's laws and the conservation laws lies in the fact that they are seen to be true by all observers, no matter how these observers are moving relative to the system being observed.

3. The Third Law of Motion

One loose thread waits to be tied: Newton's Third Law of Motion. Actually, we already have it, hidden away in Conservation of Momentum. Suppose we have two bodies (M_1 and M_2) interacting with each other. At one instant of time their velocities are v_1 and v_2, while at a later moment their velocities are v_1' and v_2'. The total momentum of the two bodies is constant for all time, so that we can write:

$$M_1 v_1 + M_2 v_2 = M_1 v_1' + M_2 v_2'$$

This equation can be rearranged to read

$$M_1(v_1' - v_1) = -M_2(v_2' - v_2)$$

But $v' - v$ is simply Δv, the change in velocity during the interval of time Δt, so that by dividing both sides of the equation by Δt we can write it as

Discovering the Natural Laws

$$M_1 \frac{\Delta v_1}{\Delta t} = -M_2 \frac{\Delta v_2}{\Delta t}$$

Now we recognize that $\Delta v/\Delta t$ is nothing more than the acceleration of a moving object, so that the equation reduces to

$$M_1 a_1 = -M_2 a_2$$

But if M_1 is experiencing an acceleration a_1 there must be a force F_1 acting on it, and similarly there must be a force F_2 acting on the other mass. Thus we have proven that

$$F_1 = -F_2$$

In other words, when two objects interact, the force acting on one must be equal to the force acting on the other, and must be aimed in the opposite direction. This statement, of course, is nothing more than Newton's Third Law of Motion.

Since Conservation of Momentum is a completely universal law, we might guess that Newton's Third Law is equally universal, applying to the interactions between all pairs of objects. Strangely enough, when we go to test this law we find that there are certain exceptional situations where the law appears to be violated. We will examine one of these situations in the next section, and we will find that the Third Law is, in fact, not a universal law, and that it does not apply to all the kinds of interactions that there are. However, it does apply to all of the forces within Newton's domain of knowledge: the gravitational force and mechanical contact forces such as occur between bouncing balls, springs, ropes, and pulleys. As for the exceptions, these turn out to be sleight-of-hand situations where an interaction that appears to take place between only two objects, as far as the naked eye can see, actually involves more than two objects.

As we will see in the next section, the above example is only one of many situations where we find, when we do careful experiments, that the classical laws of physics apply only within a very restricted domain. They are good only as long

Definitions, Experiments, and the Laws of Motion

as we don't deal with things that are too big, too small, moving too fast, or carrying electric charges.

4. Experimental Tests of Newton's Laws

It is clear from what we have said that the conclusions you draw from an experiment depend on the particular theoretical ideas you begin with—the particular conceptual structure you have erected in your mind. We can see how this logic works by looking at some of the simple experiments done in the teaching laboratory with springs and masses.

Suppose we take a kilogram mass, accelerate it by pulling on it with a spring, and measure both the acceleration and the spring's stretch. If, for example, the acceleration is one meter/sec^2, and the spring stretches a distance of 5 cm, then we know that a 5 cm stretch of the spring corresponds to a force of one Newton (one Newton being defined as the force required to give a one-kilogram mass an acceleration of one meter/sec^2). The spring is now calibrated. Now we apply the same spring to another known mass and measure the resulting acceleration.

From this experiment we can come to two different conclusions, depending on whether we are using the Newtonian approach or the Machian approach. According to the first, once the spring is calibrated we know how much force it applies, and so we can use it on a number of different masses, measuring the acceleration of each mass. In this way we show, experimentally, that the acceleration is inversely proportional to the mass. (In experiments of this type we often make some hidden assumptions about the known masses. Either we weigh them—which assumes that gravitational mass equals inertial mass—or else we say that masses add linearly, so that we can build up large masses by adding identical objects.)

Now look at the same experiment from the Machian point of view. The masses being accelerated are known by inertia-balance measurements. If different masses are accelerated by the spring, we find that the 5-cm stretch of the

Discovering the Natural Laws

spring always results in the same product $F = ma = 1$ Newton. It follows, as before, that the acceleration is inversely proportional to the mass. In addition the experiment shows that the force exerted by the spring (for a given stretch) remains the same in all the trials regardless of the type of mass or its velocity, facts that had to be assumed in the other interpretation.

In another experiment we add a second spring in parallel with the first (Figure 2-6). The Newtonian experimenter

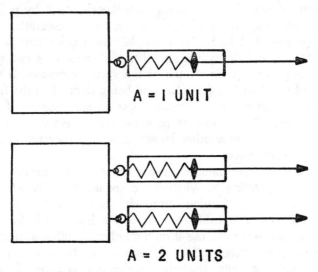

Figure 2-6. The Newtonian observer says: Two springs give twice as much force as one spring. Therefore there is twice as much acceleration, as verified by this experiment. The Machian observer says: Two springs produce twice as much acceleration as one spring. This requires twice as much force, by definition, and so the experiment demonstrates that springs add their effects linearly.

would say that this experiment shows that doubling the force doubles the acceleration. The Machian experimenter, on the other hand, notes that as a result of adding a second spring the acceleration has doubled. Therefore the force must have

Definitions, Experiments, and the Laws of Motion

doubled. The experiment thus verifies the fact that forces exerted by two springs add linearly.

We see from these examples that the conclusions drawn from an experiment depend very much on the point of view of the experimenter. We also see that in order to design an experiment to verify a law of nature we must be very clear in our minds what parts of the law are definitions and what parts remain free to be verified by observing nature.

Both the Newtonian and Machian points of view are equivalent as far as practical applications are concerned. However, there is an important conceptual difference. The Newtonian method assumes you know all the forces and all you want to do is to find the response of objects to these forces. The Machian method leaves the door open for the discovery of new forces, and in particular allows for the possibility that "fictitious" forces may turn out to be real forces. The unresolved question of inertial forces has led D. W. Sciama to say: "Newton's laws of motion are logically incomplete by themselves, and the problems they raise lead one step by step to the full complexity of General Relativity."[14]

Of course, it is well known that Newton's laws of motion are only approximations to the laws that actually govern the motion of the various objects in the universe. They are extremely good approximations as long as we deal with things that are not too big, not too small, not moving too rapidly. For objects the size of an atom or smaller, the laws of quantum mechanics take over, while objects as massive as stars require the principles of general relativity to describe their behavior accurately. Things traveling at great speeds do not obey Newton's laws in their simple form, but require the modifications of Einstein's Special Theory of Relativity.

For these reasons nobody in this day and age sets out to do experiments verifying Newton's laws of motion to a very great degree of accuracy. The reason is that when you make the experiments sufficiently accurate, or if you do the experiments with atomic-size particles, you find that you must begin dealing with relativistic mechanics, or quantum mechanics, or both.

For example, suppose you do an experiment in which a billiard ball makes a collision with another billiard ball origi-

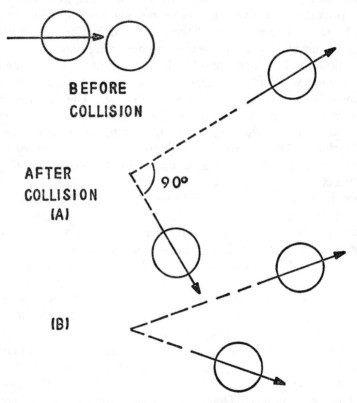

Figure 2-7. A stationary sphere is struck, slightly off-center, by a rapidly moving sphere. According to classical Newtonian mechanics, the two spheres of equal mass should move away from each other after collision along paths that are at right angles to each other. However, if the incident sphere is traveling fast enough so that relativity enters the picture and its mass is appreciably increased, then the angle between the two paths after the collision will be less than ninety degrees. Measurement of this angle is one way of verifying the mass-increase formula.

Definitions, Experiments, and the Laws of Motion

nally at rest. Since the collision will generally be a glancing blow, the two balls will go off in different directions. According to the conservation laws of Newtonian mechanics, the angle between the paths of the two balls will always be 90° (Figure 2-7). However, if you do the same experiment using fast electrons shooting through a cloud chamber, you get a different result: the angle after the collision will be less than 90°, and in fact this angle will turn out to depend on how fast the incident electron was moving.

What does this result do to conservation of momentum? In order to keep the law intact it is necessary to say that the momentum of the incident electron is greater than what you would expect from the simple expression mv, where m is the constant rest-mass of the electron. We then interpret this phenomenon by redefining the electron mass—by saying that the rapidly moving electron has a mass that is greater than its rest-mass, and that this "relativistic mass" (M) varies with the velocity according to the formula

$$M = \frac{m}{\sqrt{1 - \left(\frac{v}{c}\right)^2}}$$

(This means that in an inertia balance type of experiment, if M_1 is a known mass, then the unknown mass is defined by the relationship $M_2 = M_1 |v_1/v_2|$ where now M_1 and M_2 are the relativistic masses defined above.) With this new definition of mass, conservation of momentum remains a valid law.

This re-definition of the mass also serves to keep the form of Newton's Second Law of Motion intact. Imagine an experiment in which an electron is being accelerated by a constant electric field. We notice experimentally that the velocity does not increase at a constant rate, but that it gradually levels off until it approaches a constant value: the speed of light. In other words, the acceleration decreases as the velocity increases. (Figure 2-8) Does this mean that the amount of

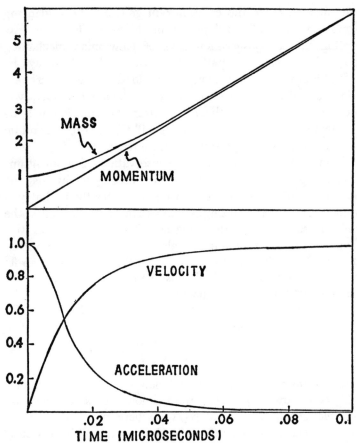

Figure 2-8. An electron under the influence of a strong electric field (one hundred thousand volts per meter) has its momentum increased at a constant rate, as shown by the straight line. However, its velocity ceases to increase as it approaches the limiting value of the speed of light, and its acceleration decreases toward zero. After some time the increase of momentum is due almost entirely to the increase of mass, while the velocity remains constant. The quantities plotted are: v/c (where v is the electron velocity, and c is the speed of light), a/a_0 (where a is the electron's acceleration and a_0 is its initial acceleration), m/M (where m is the relativistic mass and M is the rest-mass), while the momentum is plotted as the ratio mv/Mc.

Definitions, Experiments, and the Laws of Motion

force is decreasing as the electron speeds up? This is not an impossibility, for we certainly know that there are forces (such as magnetic forces) which do depend on the velocity of the moving charge.

It turns out, interestingly enough, that by allowing the electron's mass to vary according to our new definition, we find that a constant electric field produces constant force, even though the electron's acceleration varies. For, we recall that by definition, the force is the rate of change of momentum of the electron. The experiments show that the momentum increases at a constant rate, indicating a constant force. At the same time, if the electron mass increases while the particle is speeding up, the acceleration can decrease to zero even while the force remains constant.

When we apply Newton's laws of motion to the force between two electrically charged objects, we must be extremely careful, for the force involved is an extremely complicated one. In elementary books on electricity we learn that there is an electric force (either an attraction or a repulsion) between two charged objects, and that the amount of force depends only on the distance between the objects—not on how fast they are moving. When we get a little more advanced, we learn that if the two charges are in motion, then there appears a new component of force that is directed at right angles to the motion of the objects, and we call that a magnetic force. As long as the magnetic force is the result of electrons moving while confined within copper wires, the problem is not too complicated.

But just let us think of the simplest situation in nature: two electrons moving past each other in some arbitrary direction. Immediately we open the lid on a Pandora's box of the most paradoxical complications. In Figure 2-9 we show two electrons moving in paths directed at right angles to each other. (Notice that the velocities are shown relative to some outside observer.) Let us assume that for the time being they are moving with constant speed—thus each of them is producing both an electric and a magnetic field. (Actually there

Discovering the Natural Laws

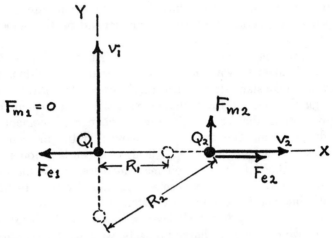

Figure 2-9. Diagram illustrating the fact that the electromagnetic forces acting on two moving charged particles are neither equal in magnitude nor opposite in direction. Q_1 is an electric charge moving with velocity v_1 along the y-axis. Q_2 is an electric charge moving along the x-axis. The black circles show where the charges are located at the present instant of time. One of the dotted circles shows where Q_1 was at that earlier instant of time when it emitted the electromagnetic field presently felt by Q_2. R_2 is the distance that field had to travel. The dotted Q_2 and the distance R_1 have similar interpretations. Q_1 feels an electric force F_{e1}, but feels no magnetic force. Q_2 feels an electric force F_{e2} and also a magnetic force F_{m2} directed in the y direction. Notice that even though F_{e1} and F_{e2} are in opposite directions (by definition) they are not equal in magnitude. The magnetic forces are clearly not equal in magnitude. The total force on Q_2 is the vector sum of the electric and magnetic forces acting on it, and this force is neither equal to nor in the direction opposite to the total force acting on Q_1.

is only one electromagnetic field, but we find it convenient to break the force up into two components: the electric force is directed along the straight line between the two particles, while the magnetic force is perpendicular to this line.) Since the two particles are moving past each other, each particle experiences an electric and magnetic field that is continually

Definitions, Experiments, and the Laws of Motion

changing. The changing magnetic field, of course, contributes to the electric field. In addition, it takes a certain amount of time for these changes in the field to travel from one electron to the other. Notice that the diagram shows the two electrons in black at one instant of time, but there is also shown (dotted) the position of the particles at an earlier instant of time. The dotted position shows where Q_1 was when it originated the electromagnetic field that finally reached Q_2 at a later time. The distance the field had to travel is labeled R_2. Similarly, when Q_2 was at its dotted position it emitted the field that later reached Q_1. Since the distance R_1 is not the same as R_2, there is no reason to expect the forces acting on the two particles to be equal.

This is, in fact, the case. If you carefully calculate the instantaneous forces acting on the two charges, and ask whether they are equal and opposite, you find that neither the electric part of the force nor the magnetic part of the force are equal.[15] (There are complications, which I have not mentioned here, caused by the way in which the electric and magnetic fields are transformed by the equations of relativity.) As a result, the total force acting on Q_1—adding up both electric and magnetic parts—is neither equal in magnitude to the force acting on Q_2, nor is it aimed in the opposite direction.

This observation is always a shocking one the first time it is encountered, for we have been taught for so long that according to Newton's Third Law of Motion the mutual forces between two objects are equal in magnitude and opposite in direction. Yet here we find that the simple interaction between two electric charges does not obey the law. Is the Third Law not a universal law?

What is more embarrassing is the fact that earlier in this chapter we derived Newton's Third Law by deduction from conservation of momentum, which we claimed to be completely universal. If conservation of momentum is universal, why is not Newton's Third? Where did we go wrong?

When we investigate the situation carefully, we find that

43

the story as we have told it above is not the entire story. In order for the electrons to travel in straight lines as we have described, their paths must be restricted by forces coming from the outside, so that we would no longer have a simple interaction between two isolated bodies. If, in fact, these two electrons are all alone, then they cannot travel in straight lines, because their mutual forces will cause them to accelerate and travel in some sort of curved path. As a result of this acceleration they must give off electromagnetic radiation—or photons, if you prefer. Thus we do not have a simple system consisting of two isolated electrons, but we have a complicated system that also includes the surrounding electromagnetic field, or its equivalent: the photons flying back and forth between the electrons and going off in all directions. If all the parts of the system are taken into account, momentum is strictly conserved as the two electrons move past each other; we just have to make sure that we include the momentum of the electromagnetic field in the calculations. As for Newton's Third Law, it really doesn't apply, for we are not dealing with a simple two-body system with a single force-pair acting between them. (The situation is different if we have electric currents passing through wires; the currents must make closed loops, and if we add up the magnetic force acting on all parts of the loops, we find that the Third Law does apply to the total force between the two loops.)

The point of the above discussion is that when we deal with atomic-size systems or when we deal with moving electric charges, we quickly get into areas where the simple forms of Newton's laws are not even approximately true. Thus, modern experiments on the fundamentals of mechanics mainly deal with questions involving Einstein's theory of relativity, and it was in the development of this theory that Einstein (and many others) really came to grips with the underlying definitions and assumptions of physics.

5. *Gravitation and the Laws of Motion*

One area where Newton's laws in their classical form might

Definitions, Experiments, and the Laws of Motion

be expected to apply with complete accuracy is in problems concerning the motion of planets and space vehicles. The very fact that we can guide ships through space with the precision required to land on a small area of the moon shows that our knowledge of the laws of motion is complete enough for all practical purposes. But even here we cannot point to one special group of experiments designed to test the assumptions of the second law of motion all by itself, for this law of motion is always an integral part of the equation that predicts where a space vehicle is going to move under the influence of the gravitational force guiding it. Any observation of planetary or satellite orbits automatically tests the second law together with the law of gravitation. The two are inseparable.

For notice that in any real situation, what we actually measure is the *acceleration* of the orbiting body; the strength of the gravitational force is inferred by making use of the Second Law. For example: what we learn from Kepler's Third Law of motion is a relationship between the radius of a planet's orbit and the time it takes for it to go around that orbit. This is an observed fact. We also know from geometry that an object traveling in a circle is continually accelerating toward the center, with an amount of acceleration given by $a = v^2/r$, where v is the velocity around the orbit, and r is the radius of the orbit. Since Kepler's Third Law gives us a relationship between the radius and the velocity, we can use this to eliminate the velocity from the two equations, and in this way we find that the centripetal acceleration is inversely proportional to the square of the radius: $a = -K/r^2$, where K is a number that gives the magnitude of the acceleration.

Knowing the acceleration of this body at every instant as it travels around its center of gravitation, we can find out where it is going to be and how fast it will go at any future instant of time. The position and velocity of any orbiting object are the actual observables. The notion of *force* is an added feature. It is an interpretation of what we see, expressing our feeling that there must be a cause for the observed acceleration. But it is not the only interpretation possible. Einstein's general theory of relativity is a theory of gravitation that uses the concept of motion along the lines of curved space instead of saying that the motion is caused by a force acting through the medium of a gravitational field.

If we say that the acceleration of the satellite is caused by a gravitational force, then according to the Machian point of view, the magnitude of the force is given (by definition) by the expression $F = ma$. The inverse-square law of acceleration becomes an inverse-square law of force: $F = -mK/r^2$. The point being made here is that the Second Law of Motion cannot be separated from the law of gravitation in any actual tests of these laws based on a study of orbit motion. As we shall see in Chapter 4, orbit studies are our most accurate tool for studying gravitation.

In a real situation, of course, the orbit of a satellite or planet does not depend only on the gravitational field of the body at the center of the orbit. A satellite circling the earth at an altitude of a few hundred miles is beset by perturbing tugs from the moon, the sun, and all the other planets. When we calculate the orbit of this satellite taking into account all of these perturbations, we use not only the inverse-square law, but also the fact that the forces add according to the vector rule. The law of vector addition is an important ingredient of the laws of motion that can be tested experimentally. The feature of the law that we do not test is the basic idea that the quantity of force equals the mass times the acceleration. Since this is true by definition, we are free to use it as a tool for measuring the force.

Once more, I emphasize that this particular choice of definition is not the only way of visualizing the relationship between force and motion. I have been using the Machian approach. A person using the Newtonian method would write the above paragraphs somewhat differently.

However, whichever way you choose to do it, you will find that Newton's laws of motion are so closely intertwined

Definitions, Experiments, and the Laws of Motion

with the other laws of physics, such as the conservation laws, the law of gravitation, Coulomb's law of electric force, and so on, that the verification of these laws to be described in the remaining chapters will automatically include verification of the laws of motion. The reality that exists outside our minds is independent of the choice of words or concepts that we use to translate that reality into little manageable packages of thoughts.

CHAPTER 3

The Four Forces of the Physicist

The development of the *force* concept marks a clear-cut boundary between the habits of thought of the ancient philosopher and the modern scientist. At a very early stage of history inanimate objects were believed to have internal powers and were endowed with a kind of life of their own, accounting for the actions of clouds, rivers, storms, and stones.[1] The motion of the planets through the sky was associated with gods, goddesses, and assorted supernatural powers.

Gradually the notion arose that motion was the result—not of occult powers—but of natural attractions and repulsions between material objects. An important turning point was the realization at the time of Galileo that the essential function of a force was not to produce *motion*, but rather to produce a *change* of motion. It was this important concept (summarized in Newton's Second Law) that set modern mechanics on its present course.

Once the new point of view was adopted, physics became flooded with a host of "forces." Every type of phenomenon that occurred was regarded as being caused by a force, and thus the unwary student was beset with a variety of names to learn: gravitational, electrical, magnetic, hydraulic, pneumatic, chemical, adhesive, cohesive, centrifugal, Coriolis, and capillary, among others. In one sense these forces were little different from occult forces, since their basic origin was not at all understood. There was one important difference, however: The strength and behavior of these forces could be

The Four Forces of the Physicist

measured, and so they were a means of bringing quantitative order into nature.

With the coming of the twentieth century, another important change took place in the conceptual structure of science: the idea that all matter is composed of atoms, and that these atoms in turn consist of a relatively small number of elementary particles. This concept produces a great simplification in our study of the fundamental behavior of matter. Instead of having to understand the different behavior of over one hundred separate elements and hundreds of thousands of compounds, we need only concern ourselves with the interactions between just a few particles: electrons, protons, neutrons, and photons—plus the neutrinos and assorted mesons that play a role only within the depths of the atomic nucleus.

With this kind of model to work with, we find that the fundamental laws of nature take on a new depth and simplicity. To verify the law of conservation of energy, we no longer must measure the output of electrochemical reactions, thermoelectric reactions, photoelectric reactions, and so on. We no longer worry about perpetual motion machines. Instead, what we say is this: All activities in nature occur as a result of interactions between pairs of particles. If we find that energy is conserved in all of these elementary interactions, then energy will continue to be conserved in reactions involving larger groupings of the elementary particles.

For example, a hydrogen atom is simply one electron orbiting around one proton, the force holding them together being the electrical force. As we will show in Chapter 5, energy is always conserved in any kind of reaction caused by the electric force and, in accordance with this, we find that a hydrogen atom in its normal state always has a certain definite, constant amount of internal, or binding energy. If one hydrogen atom joins with another hydrogen atom to form a molecule, the structure of this molecule is still determined by forces between pairs of particles—except now there are several pairs of forces: electron-proton, proton-proton, and electron-electron (Figure 3-1). If two hydrogen molecules

Discovering the Natural Laws

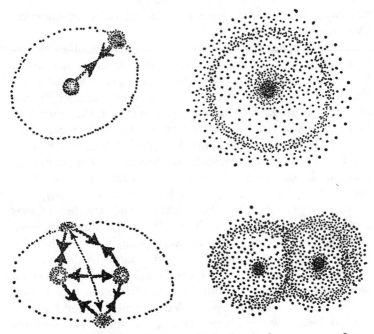

Figure 3-1. *Top:* The simplest model of a hydrogen atom shows an electron orbiting a proton, the two held together by the force of electrostatic attraction. On the right is a representation of the atom as depicted by quantum physics, in which the electron is somewhat like a cloud or wave packet surrounding the nucleus. *Bottom:* In a hydrogen molecule there are two protons and two electrons all bound together. The protons repel each other, as do the two electrons; the protons attract the electrons, so that there are altogether six force-pairs. However, it is impossible to explain how a hydrogen molecule holds together with such a model. The quantum-mechanical picture shown on the right depicts the electron cloud as swarming around both protons, so that the entire structure holds together. In this picture the idea of forces being represented by arrows between point charges no longer makes sense. Physicists now prefer to speak of "interactions" rather than "forces."

join with an oxygen molecule to form two water molecules, the water molecules as a whole have less internal energy than the original hydrogen and oxygen. What happened to the

The Four Forces of the Physicist

extra energy? We find that the newly formed water molecules are now moving around rapidly; they have more kinetic energy than before. In other words, the joining together of the two molecules has changed some of the potential, or internal binding energy of the system into a random kinetic energy of the molecules as a whole darting about within the fluid. Looking at the reaction from the outside, we say that burning hydrogen and oxygen to form water has produced heat.

In the old days when people did not know that heat energy was nothing more than our perception of the increased kinetic energy of particles down on the atomic and molecular level, it was necessary to make careful measurements to show that this heat energy was equal to a certain definite amount of mechanical energy in order to prove that energy was conserved; that is, burning a definite quantity of fuel produced enough steam to make an engine do a certain amount of work.

Nowadays on an engineering level it is still useful to talk about the heat of combustion of fuel and the mechanical equivalent of heat. But, on a more fundamental level, we know that heat is nothing more than a form of mechanical energy. The heat energy within an object is related to the kinetic energy of the molecules moving around within that object, together with the oscillatory or rotational motions taking place within the molecules, plus the potential energy of the electrical forces binding the atoms together within the molecule.

Heat energy, with this model, is not a separate, special form of energy. It is nothing more than kinetic and electrical potential energy.

Thus it is not necessary for us to investigate all the possible different kinds of chemical-electrical-thermal-mechanical reactions in order to verify the Law of Conservation of Energy. All we have to do is to show that when individual electrons, protons, and neutrons interact with each other, energy is neither gained nor lost. Since all material objects are composed of electrons, protons, and neutrons, it follows that in any interaction between material objects energy will be neither gained nor lost.

How do we account for energy in the form of light, radio waves, gamma rays, and so on? To take this kind of energy into account, all we need do is to introduce another kind of particle—the photon—the quantum of electromagnetic energy. Using the concept of the photon, we can understand reactions in which atoms emit light, as well as the converse reactions in which materials bombarded with light emit electrons. We can also understand the manner in which energy passes from the sun to the earth—it takes place by means of a steady stream of photons with wavelengths ranging over the entire spectrum, from the far radio region all the way down to the gamma ray sector.

Now we have arrived at a very simple picture of the universe: Everything is built up out of electrons, protons, and neutrons, with photons passing back and forth between them as carriers of electromagnetic energy. Since there are such a few fundamental building blocks, we might expect that there are only a few basic ways in which these particles can interact with each other. In fact, on examination we find only four different kinds of interaction between elementary particles.

In other words, instead of talking about the multitude of forces that plagued the classical physicist, the modern physicist is interested in only four basic forces: the gravitational, the electromagnetic, the strong nuclear, and the weak nuclear forces.

The gravitational force is the most widespread of all the forces. On a large scale we think of it as the force that holds objects down to the surface of the earth, the force that keeps planets in their orbits, the force that maintains the stars in their paths within the galaxy. On the microscopic level, the gravitational force is a force of attraction between all the elementary particles; it is the only one of the four forces that acts in the same manner between all the types of particles. Neutrons, protons, electrons, and photons all attract

The Four Forces of the Physicist

each other according to the same general law: The strength of the force is proportional to the product of the masses and inversely proportional to the square of the distance between them. Since the same law applies to all objects, Einstein considered gravitation to be the result of the geometrical properties of space itself.[2] In working out the consequences of this idea, Einstein found that the gravitational interaction was not a simple inverse-square law such as Newton's law, but was of a much more complex nature. However, the deviations from Newton's law were so tiny that for almost all problem-solving purposes Newton's law could be used to a very high degree of accuracy.

When we observe the interactions of electrons and protons with each other, we see that the effect of the gravitational force is completely overwhelmed by a much stronger force: the electromagnetic force. The electromagnetic force differs from the gravitational force in at least two very important ways. First of all, the electromagnetic force comes in the form of both attraction and repulsion, while the gravitational force only comes as an attraction. Electrons attract protons, but electrons repel other electrons, and protons repel protons. We describe this by saying that electrons and protons carry something called an "electric charge," and "like charges" repel, while "unlike charges" attract. The second important feature of the electromagnetic force is that it is much stronger than the gravitational force. In fact, the only reason we can notice the gravitational force at all is the fact that a normal object has in it an equal number of electrons and protons—thus we say that it is electrically neutral. When two objects are nearby, the electrical attractions and repulsions cancel each other out, and so we are then aware of the remaining small gravitational force.

The simple electrical attraction or repulsion we have been discussing is called the electrostatic force, for it is found between charged particles that are at rest. If the particles are moving, then the force becomes more complicated. For example, consider two copper wires lying parallel to each other; they are electrically neutral, so that the electrostatic forces between the electrons and protons cancel each other out. But if some of the electrons are moving from one end of the wire to the other, then a new force appears: If the electrons in both wires are moving in the same direction, the wires are attracted to each other; if they are moving in opposite directions, the wires are repelled. We say that there is a magnetic force resulting from the electric current in the wires. Actually, from the modern point of view there is only one kind of force acting—the electromagnetic force. This force has two components, one that depends simply on the distance between the electrons—the electrostatic force—and another component that depends on the velocity of the electrons as well. The latter is what we call the magnetic force, but it must be clearly understood that this is not a new and different force, but that both the electric and the magnetic forces are two components of a single interaction.[3]

Photons are not affected by electric or magnetic forces—that is, a beam of light shot through a magnetic field is not deflected as is an electron beam. But when atoms emit photons of light, it is the electromagnetic force that is responsible for this emission. Similarly, when an atom absorbs a photon of light and shoots out an electron (the photoelectric effect), the electromagnetic interaction is the cause of this event. In fact, the highly abstract theory of "quantum electrodynamics" describes the electromagnetic force existing between any two charged objects as resulting from the continuous back-and-forth interchange of photons. In this model the photon is the carrier of the force, and all the effects that we attribute to electric and magnetic fields are simply the visible results of this invisible exchange of photons (Figure 3-2).

Within the nucleus of each atom are a number of protons. If the gravitational and electromagnetic forces were the only forces available, no atomic nucleus could hold together. The

The Four Forces of the Physicist

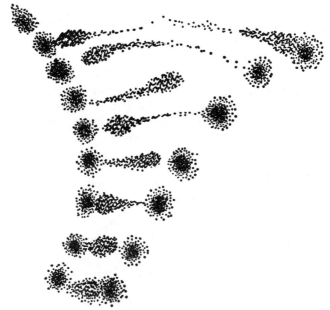

Figure 3-2. Two streams of electrons repel each other and travel in diverging paths. The theory of quantum electrodynamics pictures the repulsive force as resulting from a continual back-and-forth exchange of "virtual photons" that are emitted and absorbed by the electrons. The picture shown here is entirely imaginary and is to be considered more symbolic than literal.

protons would repel each other with an enormous amount of force, far too great for the gravitational attraction to overcome, and immediately the nucleus would blow itself apart. In order to explain how it is that atomic nuclei hold themselves together, a new type of force must be invoked: This is the "strong" nuclear force.

The strong nuclear force acts as an attraction between the heavy particles—the neutrons and the protons. As far as this attractive force is concerned, there is no difference between neutrons and protons; in fact, both these particles are generally considered to be two different forms of one type of particle, differing only in the fact that the proton has an electric charge.

A most important characteristic of the nuclear force is that it is a short-range force: It comes into play only when the particles involved are close enough so that they are essentially touching each other. If you shoot a neutron at a proton, the neutron will not interact with the proton at all unless it makes a head-on collision. If that happens, the two will suddenly fall violently together, emitting a quantum of energy in the form of a gamma ray photon (Figure 3-3).

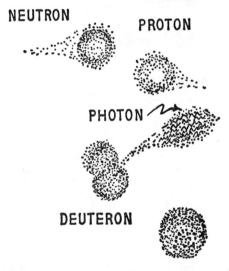

Figure 3-3. When a neutron encounters a proton under favorable conditions, the two particles may join together to form a deuteron. In doing so, a high-energy photon (a 2.2 MeV gamma ray) is emitted, so that the deuteron has less energy (and mass) than the initial neutron and proton.

The result is a stable nucleus of deuterium (heavy hydrogen), held together by the strong nuclear interaction. On the other hand, if you bring a pair of protons together they will not remain together even if they are close enough to touch.

The Four Forces of the Physicist

There must be at least one neutron together with the two protons to make a nucleus that will hold together for any length of time. In general, we always find within an atomic nucleus at least as many neutrons as there are protons. When there are very many protons inside a single nucleus, even the nuclear binding force cannot hold them all together, and the nucleus has a tendency to get rid of this extra energy by spitting out an alpha particle, a combination of two neutrons and two protons. From this point of view, it is the electromagnetic force that is responsible for radioactivity, as well as for nuclear fission. The nuclear force does its best to keep the nucleus held together.

The weak nuclear interaction is a rather mysterious force, whose action is concerned mainly with the radioactivity of the neutron. A neutron, by itself, is an unstable particle, turning into a proton plus an electron plus an antineutrino with a half-life of about twelve minutes. This twelve minutes is a very long time as far as nuclear events is concerned, and indicates that the cause of this transformation is a force of a very weak nature, known as the "weak nuclear force." When neutrons are embedded in a nucleus, the presence of the protons may stabilize the situation so that this transformation does not take place. However, there are many isotopes where the nucleus contains one or two more neutrons than necessary, and in the course of time each of these neutrons transforms itself into a proton, spitting out an electron and an antineutrino. The ejected electron we observe as a beta particle, while the antineutrino is unobservable except by means of very elaborate instrumentation. The weak nuclear force is thus responsible for all radioactive transformations involving the emission of beta radiation. There are also reactions involving mesons (for example, the transformation of a pion into a muon plus a neutrino) that are caused by the weak nuclear force.

Of the four forces, the gravitational is by far the weakest, even weaker than the "weak" nuclear force. The following table compares the strengths of the other three forces with the gravitational force:

INTERACTION	RELATIVE STRENGTH
Gravitational	1
Weak nuclear	10^{27}
Electromagnetic	10^{38}
Strong nuclear	10^{40}

What this table means is that the nuclear attraction between two protons just touching each other is a hundred times greater than the electrostatic repulsion between them. (In spite of this, two protons will not stick together for reasons having to do with the wave nature of the particles.) By contrast, the electrostatic repulsion between two protons is 10^{38} times greater than the gravitational attraction between them. For this reason the gravitational force is completely ignored when we discuss the electromagnetic and nuclear interactions of particles.

In spite of the fact that there are many mysteries about the nuclear forces, we do have a great amount of experimental evidence about the way these forces behave. That is, we can observe what happens when two particles come together under the influence of these forces, and from these observations we can deduce how the strength of the force varies with distance, and how it depends on properties of the particles such as electric charge and spin.

Now that we have identified the four fundamental forces, the first question that comes to mind is this: Are there any other forces? Are there any events observed in the universe that cannot be explained by means of the four interactions we have been discussing?

In spite of the fact that there is little discussion about it in the scientific literature, this is perhaps the most important philosophical question facing science today. If the answer is positive, if there are phenomena so unexplainable that an entirely new kind of force must be assumed to explain them—

The Four Forces of the Physicist

then the future of science is open-ended. With the new kind of force may be associated an entirely new spectrum of particles with properties and functions as yet unimagined, generating the kind of revolution in science that was caused by the realization that all the properties of matter are the result of interactions between elementary particles.

On the other hand, if the answer is negative—if the four forces are all there are—then we know all the ways that matter can behave on a fundamental level, and that's the end of it. It's not the end of science, to be sure. There are still an infinite number of questions to be answered: We still don't understand many details concerning the fundamental structure of three out of the four forces. The electromagnetic force is the only one that physicists feel has been adequately described, through the theory of quantum electrodynamics. Furthermore, as soon as we put more than two particles together, we find ourselves dealing with situations that are very difficult to analyze in an exact way, so that the practical problem of using our fundamental knowledge to predict how matter is going to behave is a task that will not be exhausted for centuries to come. Simply pass an electric current through a tube filled with hydrogen gas surrounded by a magnetic field and you will generate a host of waves, instabilities, and ionic motions that have yet to be fully described or understood—in spite of the fact that the force guiding each particle is electromagnetic and is therefore well-known. The situation is simply so complicated that it evades analysis. On top of this, astrophysics is bursting with newly observed phenomena: quasars, pulsars, magnetospheres, gravitational collapse. Biophysics has before it the fantastically complex job of determining how living organisms really work, and how thinking creatures really think. There are new and unexpected discoveries waiting to be uncovered.

Yet with all the unexplained phenomena already observed, rarely does any scientist suggest that perhaps we need a fifth kind of force to complete the theory and account for the unknown. There is a general agreement that the four familiar forces ought to account for all chemical reactions, all biology, all astrophysics, and all nuclear physics. Occasionally an article appears in one of the journals suggesting that perhaps one of the unexpected results obtained in experiments with high-energy elementary particles can be explained by the assumption of a fifth force. Always the unexplained experimental result is a very small perturbation on top of the effect that normally would be expected from the known four forces, and the fifth force suggested is an ultraweak force whose effects are so subtle that they are normally not noticed at all.

The general impression is that, if there were a fifth force lurking under cover—a strong fifth force interacting with normal matter under ordinary conditions—that force would have been noticed by now, for it would make matter behave in ways different from those required by the old four forces. Nobody seems inclined to introduce a new type of interaction to explain any phenomena having to do with the bonding of atoms to form crystals and molecules, or solid, liquid, and gaseous matter with the conglomeration of these molecules to form living organisms. The electromagnetic force, plus the laws of quantum mechanics, appear to be sufficient to account for all of these activities.

In the remainder of this book we will examine some of the methods scientists use to observe the behavior of particles under the influence of these four forces. We will not be concerned with the minute details of what happens when two or more particles come close to each other, for that is a very large subject that covers the area of elementary particle physics, nuclear physics, and atomic physics. However, we will be interested in certain very broad laws that govern the behavior of all matter, and we will want to see what is the experimental evidence that leads us to believe that each of these laws applies equally well to the four types of forces. Such laws are the conservation laws: conservation of energy and of momentum. It is not to be taken for granted that these laws are true for all the forces but, on the contrary, the laws must be taken as hypotheses and we must test each

The Four Forces of the Physicist

one of the four interactions to see if the conservation laws are good under all circumstances.

We will also want to test each of the four forces to see how the various particles respond to them. We know that neutrons, protons, and electrons react differently to an electric force. How do they respond to the gravitational force? The answers to this question will be of great importance, when, in Chapter 9, we ask ourselves how we can manage to control the gravitational force.

The guiding principle behind all of this investigation is the idea that all particles of a given type are identical. That is, all electrons have the same mass and respond to electromagnetic forces in the same way. You cannot tell them apart. Similarly with protons, neutrons, and the rest of the particles. Were it not for this fortunate fact, physics would be far more complicated than it is. It is bad enough that in a piece of silicon all the electrons have a different amount of energy. What if they had different amounts of mass?

To a certain degree this uniformity is a matter of definition. That is, we define all particles with a certain mass and electric charge to be electrons. If a particle comes along with a different set of properties, we call it another particle. For example, the muon (one of the several kinds of mesons) is identical in almost all ways with the electron, except that it has a mass about 207 times greater than the electron. What makes it possible for us to talk about different kinds of particles is the fact that the masses and electric charges of the particles are *quantized*. That is, instead of particles coming in all sizes, they come in a few very sharply defined groups.

The fact that all electrons are identical may seem like an obvious truism; yet there is no theoretical reason requiring nature to behave that way. Such a reason may be discovered in the future. In the meantime, the knowledge that each type of particle "does its own thing" will turn out to be extremely important in Chapter 9, when we ask questions about the types of phenomena that are allowed or forbidden.

CHAPTER 4

The Gravitational Force

1. *Gravitational and Inertial Mass*

Newton's Universal Law of Gravitation is deceptively simple. In its usual form it can be stated this way: If we have two objects whose masses are M and m, and if the distance between their centers is r, then the amount of gravitational force between these bodies is proportional to the product of the masses, and inversely proportional to the square of the distance. That is:

$$F = -\frac{GMm}{r^2}$$

In this formula, G represents the universal gravitational constant, developed experimentally as a refinement of Newton's law by Henry Cavendish. At first glance, G appears to be nothing more than a constant of proportionality, but in fact G is a quantity that tells us how strong the force is, and there is no reason to assume ahead of time that G is actually a constant. How do we know that G does not depend on the kind of materials the two objects are made of? Is the force between two kilograms of aluminum exactly the same as the force between two kilograms of gold? Furthermore, does G always remain the same as time passes?

The minus sign is another important part of the formula. What it means is that two objects always *attract* each other with the gravitational force. A plus sign would stand for a repulsion. Here is another fact that must not be taken for

The Gravitational Force

granted. True, all the things we are familiar with do fall down —we would think it terribly strange if we saw something fall up (and yet science fiction writers like to talk about antigravity). How do we know that this force is always an attraction? Is it possible that under some very unusual set of conditions two objects might repel each other? Is it possible that an object made of antimatter (positrons and antiprotons) would fall upward instead of downward? These are questions that cannot be answered by theory, but must be settled by experiment and observation.[1]

Another important feature of the gravitational formula is that the amount of force depends on the magnitude of these two quantities m and M that we call the masses. If you think about this fact carefully, you will realize that this is a rather strange thing, for the mass of an object was defined in Chapter 2 as being a measure of the *inertia* of an object—its resistance to being accelerated. Now in the Law of Gravitation we treat the mass as somehow being the *source* of the gravitational field—just as the charge on an electron or proton is the source of the electrical field surrounding it.

Since we have no way of knowing beforehand that there is a connection between the inertia of an object and the gravitational force it feels, we must start out by allowing for the possibility that they are two different things. Therefore, we call one of these properties the *inertial mass* (m_i) and we call the other the *gravitational mass* (m_g). We now ask what experiments can be done to find out how these two quantities are related to each other.

2. Falling Objects

One of the first things that scientists found when they began to abandon the viewpoint of the Middle Ages was the fact that objects of different weights and materials fall under the influence of gravity with the same amount of acceleration. Although Galileo is reputed to have performed this experiment for the first time, using the Leaning Tower of Pisa as a laboratory, more accurate reports indicate that probably

Discovering the Natural Laws

Galileo never did this particular experiment. In fact, a Dutch military engineer named Simon Stevin performed the experiment of dropping two pieces of lead, one weighing ten pounds, the other one pound, from a height of thirty feet. When they hit the wooden floor, the sound of the impact of the two weights could not be separated. This opening blow in experimental dynamics was accomplished in 1586, two years before Galileo became professor at the University of Pisa. Further experiments convinced Galileo that all objects, regardless of composition, responded to the pull of gravity with the same acceleration, as long as one managed to eliminate the influence of air resistance. Until it became possible to set this observation within the theoretical framework of Newton's Second Law of Motion, its full importance could not be appreciated. Newton's law tells us that if an object with inertial mass m_i is falling to the earth with an acceleration a, the gravitational force acting on it is $F = m_i a$. If we set this equal to the force given by Newton's Law of Gravitation, we find that

$$F = m_i a = -\frac{GMm_g}{r^2}$$

On the right side of this equation we have used the gravitational mass m_g because it is this property of the falling object that determines the strength of the force. To find out what the acceleration is, we divide both sides of the equation by m_i:

$$a = -\frac{GM}{r^2} \cdot \frac{m_g}{m_i}$$

The number G is a constant (more about this later), the mass of the earth M is a constant, and at a given distance from the center of the earth r is a constant. Therefore, if the acceleration a is a constant number that describes how all objects fall under the influences of gravity—and this is the fact that was found by all the experiments—then the ratio m_g/m_i is the same constant number for all falling objects.

The Gravitational Force

Now the inertial mass m_i is measured as described in Chapter 2 by comparison with a standard kilogram on an inertia balance. The gravitational mass m_g can be measured by comparing the object with the *same* standard kilogram using an ordinary beam balance or spring balance. This procedure implies that we are defining the inertial mass and the gravitational mass of the standard kilogram to be exactly equal—we are allowed to do this because we are free to define a unit of mass in any system. As a result the ratio $m_g/m_i = 1$ for the standard kilogram. But since this ratio is the same for *all* objects, it follows that $m_g = m_i$ for every object. In other words, the gravitational mass of every thing is equal to its inertial mass.

If this conclusion is exactly true for all things in the universe, it is a cause for great wonder, since there is no apparent reason for the property of inertia and the force of gravity to be so intimately related. Albert Einstein's wonderment over this situation led him through his Principle of Equivalence directly to his General Theory of Relativity, in which he considered the falling of an object in a gravitational field to be connected to the geometry of space itself. In his theory it is perfectly normal for all things to have the same gravitational acceleration—for the acceleration is a property of the space itself, not of the object, while the force is a concept that we tack onto the situation in our minds.[2]

Because of the theoretical importance we attach to the fact that all objects fall with the same acceleration, experiments to make sure that this is really so have been performed periodically during the past 350 years. Newton himself measured the periods of pendulums constructed of various materials, for the period of oscillation of a pendulum depends on the force of gravity. For this reason pendulums of various kinds have traditionally been the most common and the most sensitive types of devices used for measuring the strength of the gravitational field. Until recently, the most accurate gravity measurements were those performed by the Baron Roland von Eötvös, of Hungary, an expert in the measurement of gravitational fields by means of pendulums.[3] His measurements showed that objects of differing materials responded identically to the force of gravity, with a possible error of five parts out of a billion. That is, the uncertainties in the measurement were such that if one of the objects experienced one unit of acceleration, the other might have had an acceleration of 1.000000005 units without being detected in the midst of the disturbing influences. The difficulty of making such an accurate observation is something that cannot be appreciated until it is tried. The slightest difference in temperature, the faintest wisp of an air current—even the gravitational force from the experimenter standing near the apparatus—is enough to produce errors big enough to spoil the experiment.

Beginning in 1959, a group at the Princeton University physics department led by Robert H. Dicke performed an updated version of the Eötvös experiment, using some of the more refined techniques of modern electronics to extend

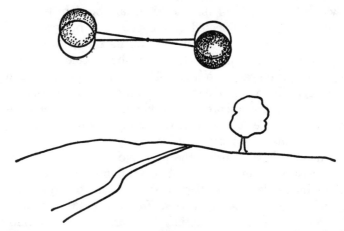

Figure 4-1. Two weights on the end of a rod are falling freely. If the one on the right falls with a greater acceleration than the other, the rod will tend to twist about a pivot in the center, and this twist could be detected by an observer falling with the apparatus.

The Gravitational Force

the accuracy of the measurement by two or three more decimal places. The principle used was not to measure the acceleration of a falling object directly. In order to do this by timing a weight falling from the Leaning Tower of Pisa, one would have to measure the time within a hundred millionth of a second in order to duplicate Eötvös' accuracy, and Dicke intended to do better. The method he used was a modification of the Eötvös pendulum, but instead of measuring the force of the earth's gravitational field, Dicke observed the response of the pendulum to the sun's gravitational field.

This appears to be a strange approach, but there is a good reason for it, as we will soon see. To understand the experiment we first think of two weights falling together, connected to each other by a light rod. If the two weights fall at exactly the same rate, the rod will remain horizontal. If one of the weights falls with a greater acceleration than the other, then the rod will twist about a pivot through its center (Figure 4-1). Now fasten a string to this pivot and assume that the rod with the two weights hangs over the earth's North Pole (Figure 4-2). (Actually, the experiment was done in Princeton, New Jersey, but it is easier to visualize the motions if we put it at the North Pole.) This arrangement forms a torsion balance. While the weights hang down toward the center of the earth, the earth and the entire apparatus are circling in an orbit around the sun. This means that the two weights are falling toward the sun, just as any objects in free orbit are actually accelerating toward the center of gravity—it is their lateral motion that prevents them from falling all the way in.

Suppose that one of the weights (the darker and denser one) happened to have a greater acceleration toward the sun than the other. This motion would cause the rod to twist the suspending wire. At the same time the earth is rotating on its axis, once every twenty-four hours, and the weights are rotating along with it. The extra acceleration of the darker weight makes the torsion pendulum rotate a little bit faster than the earth. After twelve hours of rotation, the earth and

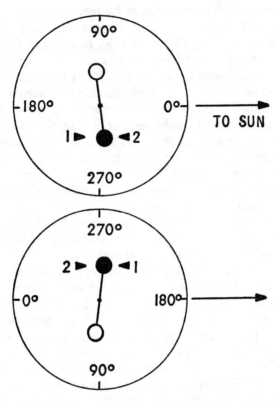

Figure 4-2. Two weights fastened to the ends of a rod are pivoted above the North Pole of the earth. The earth and everything on it is orbiting the sun, and therefore is accelerating toward the sun just as in free fall. If the black weight accelerated toward the sun more than the white weight, it would twist the pendulum toward marker 2. However, as the earth turns on its axis and the pendulum turns with it, after twelve hours have elapsed the black weight now tends to fall toward marker 1. The observer on the earth sees the pendulum oscillate back and forth with a twenty-four-hour period.

the pendulum have turned around through an angle of 180°, so that now the denser weight, accelerating toward the sun faster than the other weight, tends to swing the torsion pendulum back in the other direction. The result is that the tiny

The Gravitational Force

difference in the gravitational acceleration of the two weights causes the torsion pendulum, as seen from the earth, to oscillate around its pivot point with a period of twenty-four hours.

Such oscillating motion is easier to detect than a deflection of the pendulum in one direction only. For this reason the idea of using the attraction of the sun rather than the earth was an essential part of a very clever scheme that took advantage of every electronic trick to improve the signal-to-noise ratio of the system. The actual apparatus used three weights, two of aluminum and one of gold (an earlier version of this experiment used copper and lead). These weights were hung in a triangular arrangement in order to reduce the effects of local nonuniformities in the earth's gravitational field[4] (Figure 4-3).

Instead of actually allowing the pendulum to swing, it was forced to remain in one position by applying a small electric field to one of the weights. A mirror attached to the pendulum reflected light from a small bulb to a very sensitive photocell. Before the light entered the photocell, it was intercepted by a wire forced to oscillate at a frequency of three thousand cycles per second, thus modulating the light signal at that frequency. If the pendulum moved from its central position, the amount of light passing the vibrating wire changed, and this change was detected by a very sensitive amplifier tuned to 3000 cps. The output of this amplifier was sent to the balancing electrodes, while a strip chart recorder kept a record of the voltage necessary to keep the pendulum in the same position.

With this system it was possible to detect a change of position of the pendulum as small as 10^{-9} radian. The result of a series of measurements failed to show a twenty-four-hour oscillation of the pendulum. The final conclusion was that both the aluminum and the gold weights fell toward the sun with the same acceleration, with a possible error of one part out of 10^{11}. In this way the results of Baron von Eötvös were improved by a factor of five hundred.

Even though the experiment was performed with aluminum

Discovering the Natural Laws

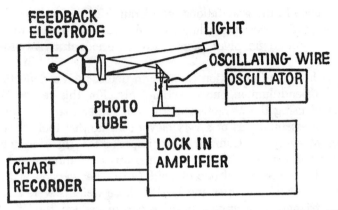

Figure 4-3. A simplified diagram of Dicke's torsion pendulum. The two white circles represent aluminum weights, and the black circle is a gold weight. A beam of light is reflected from the mirror mounted on the pendulum and is focused on the oscillating wire. If the pendulum turns, the beam grazes the wire and is modulated at the wire's oscillation frequency. The photocell picks up this pulsating light beam and sends a signal to a lock-in amplifier synchronized with the oscillator. (This arrangement has a sensitivity greater than that of an ordinary D.C. or A.C. amplifier.) The amplifier sends a voltage to the feedback electrodes, which tends to pull the pendulum back to its normal position. The strength of the feedback voltage is recorded. In this way the pendulum is not actually allowed to swing, but the voltage required to keep it from swinging is a sensitive measure of the forces acting on it. The chart record is then surveyed to find a pendulum motion with a period of twenty-four hours.

and gold (or copper and lead), we can draw a number of far-reaching conclusions from it. First, we notice that each aluminum atom contains 13 protons and 14 neutrons, while each gold atom contains 79 protons with 118 neutrons. In other words, the ratio of neutrons to protons in the aluminum is 1.07, while in gold it is 1.5. In spite of this difference in the relative number of neutrons and protons, both materials fall with the same acceleration. This evidence leads us to believe that neutrons and protons separately fall with the same acceleration, and furthermore this property is unaffected

The Gravitational Force

by the energy of the binding force within the nucleus, nor is it affected by the motions of the particles within the nucleus.

Since all the bodies that we know of are composed of neutrons, protons, and electrons, it follows that all normal things, whether composed of hydrogen or uranium, respond to the force of gravitation in the same manner. When it comes to materials made of other particles—for example, antimatter composed of positrons and antiprotons—then it is not completely safe to extrapolate into the unknown, for we know very little of the gravitational properties of antimatter.

3. The Inverse-Square Law of Gravitation

One of the chief features of Newton's Law of Gravitation —a feature that appeals to our desire for simplicity—is that the strength of the gravitational field is inversely proportional to the square of the distance between the gravitating bodies. There are a number of methods for measuring the gravitational field strength. We can measure the oscillation frequency of a sensitive pendulum, and from this frequency obtain the local acceleration of gravity "g," which is proportional to the gravitational field strength. (The frequency of a pendulum is given by the relationship

$$f = \frac{1}{2\pi} \sqrt{\frac{g}{L}}$$

where L is the length of the pendulum from the point of suspension to the center of the bob.)

Thousands of measurements of this kind have been made to determine how the gravitational field varies from place to place on the surface of the earth. After corrections are made for variations in altitude and the centrifugal force due to the earth's rotation, it is found that the strength of the earth's gravitational field is not really uniform, as you would expect if the earth were a perfect sphere. To begin with, the earth is not a sphere at all, but is an oblate spheroid—it has somewhat the shape of a fat pumpkin, due to its daily rotation. That is, if the earth were not rotating, the symmetrical gravitational force would pull the earth together into a spherical shape but, since it is rotating, the centrifugal force tends to spread the earth out at the equator.

However, even making allowances for this shape, there are still slight depressions and protrusions, possibly caused by the slow swirling and flowing of the molten material within the earth's mantle. As a result the actual form of the earth's gravitational field is not a simple inverse-square law, but is something much more complicated.

In recent years the form of the earth's field has been measured to a high degree of accuracy by observations of the orbits of satellites. A satellite in a circular orbit feels the force of gravity only at one altitude. However, the orbit will remain perfectly circular only if the gravitational force is perfectly uniform. If the force is a little weaker over one part of the globe than another, the satellite will swing out a little farther when it passes over that part of the earth, like a racing car swinging wide if it is traveling too fast for the curvature of the track.

The situation becomes even more complicated if the satellite is in an elliptical orbit, as are most satellites—both natural and artificial. The satellite, as it dips in and out of the gravitational field, samples the strength of the field at various altitudes. At each instant of time, the satellite feels the gravitational force pulling on it with a certain strength and in a certain direction. The acceleration of the satellite, according to Newton's Second Law of Motion, is in the direction of the force and is proportional to the amount of force. This acceleration determines the change in velocity during the next small interval of time. The velocity, in turn, determines how far the satellite will move during that interval of time.

By means of a high-speed computer, a scientist can use this kind of arithmetic to calculate where a satellite will move, from point to point, under the urgings of a gravitational force, no matter how complicated. Turning the problem around: If we observe where a satellite is moving, if we

The Gravitational Force

measure the shape of its path in its motion around the earth, then we can determine what kind of gravitational field caused the satellite to move that way.

We have all been taught that a satellite moves in an elliptical path under the influence of the inverse-square gravitational field. We have also been taught that the earth (or other center of attraction) is at one of the focal points of this ellipse, and that the plane of this ellipse remains fixed in space. (When we say "fixed in space" we mean that it does not move relative to the distant stars or galaxies.) What is most important is the fact that the major axis (the line passing through the two focal points and the two ends of the ellipse) also remains fixed in space (Figure 4-4). In other

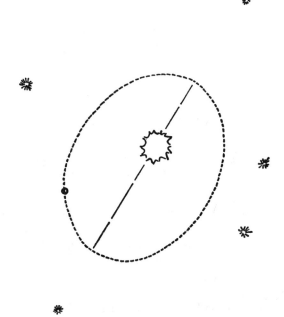

Figure 4-4. A planetary orbit in a perfect inverse-square gravitational field. The shape of the orbit is a perfect ellipse, and the major axis remains in a fixed position relative to the distant stars.

words, once the satellite is started in its orbit, the path it follows is a geometrical shape that forever remains in one orientation with respect to the distant stars.

What we have not been told in the average physics book is the interesting fact that a satellite moves in such a simple, closed, stationary elliptical orbit *only* if the gravitational field guiding it is a perfect inverse-square field. If the gravitational field differs just the slightest amount from the inverse-square law, the axis of the ellipse makes a slow rotation in space, as shown in Figure 4-5. In other words, every time

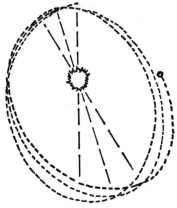

Figure 4-5. A planetary orbit in a gravitational field distorted by the shape of the sun, by the presence of other planets, and by the deviation of the sun's gravitational field from Newton's Law. The axis of the ellipse slowly swings around—or precesses—instead of remaining fixed in one position.

the satellite makes one complete orbit, instead of coming back to exactly the place where it started out—as it would with a perfect inverse-square law—it comes back to a slightly different place.

Astronomers usually locate the axis of the ellipse by finding the point in space where the satellite is closest to the central body. This point is called the *perigee* when the satellite orbits the earth, and is called the *perihelion* when we speak of objects orbiting the sun. We can then describe the rota-

The Gravitational Force

tion of the orbital ellipse by speaking of the "precession of the perihelion"—the slow motion of the orbit's bottom point.

Observation of the precession of an orbit's perihelion (or perigee, as the case may be) is thus seen to be the most sensitive test of the inverse-square law. This is one of the many reasons for the accurate measurements of planetary orbits carefully carried out by astronomers for centuries. In recent decades a great deal of effort has gone into observing the paths of artificial satellites projected into orbits of various degrees of elongation. By optical and radar methods these orbits can be measured with great precision.

Indeed, so good are the measurements that it now becomes necessary to make corrections for small forces due to sunlight hitting the satellite as well as the drag of the tenuous atmosphere reaching out into space. In addition, if the satellite becomes electrically charged due to electrons or ions or X-rays from the sun or from intergalactic space, then there will be a slight force acting on it due to the charged satellite moving in the earth's magnetic field. Also, most important are the corrections required to take into account the perturbations due to the gravitational forces from the moon, the sun, and the other bodies of the solar system. Corrections for these effects can be programmed into the computer, but there is always some uncertainty in the amount of correction to make. These uncertainties are among the main limitations to the accuracy of the measurements.

After subtracting out all the extraneous influences, we can write an equation that represents the form of the earth's gravitational field—that is, it describes how the force varies according to the location of the satellite or other observer. A simplified version of this equation is as follows:[5]

$$F = -\frac{GMm}{r^2}\left[1 - A\left(\frac{R}{r}\right)^2 P_2 - B\left(\frac{R}{r}\right)^4 P_4\right],$$

where F is the force of attraction between the earth and a satellite,

G is the universal gravitational constant
   (6.673×10^{-11} m^{3}kg^{-1}sec^{-2}),

M is the mass of the earth,
m is the mass of the satellite,
r is the distance from the center of the earth to the satellite,
R is the radius of the earth at the equator,
$A = 2.16540 \times 10^{-3}$,
$B = 7.36 \times 10^{-6}$,
$P_2 = (3 \sin^2\psi - 1)/2$
$P_4 = (35 \sin\psi - 30 \sin^2\psi + 3)/8$
and ψ is the latitude of the satellite's position.

There are a number of important features of this equation. We notice, first of all, that Newton's simple inverse-square Law of Gravitation is just the first term—and the largest term—and is followed by other quantities involving the fourth and sixth powers of the distance. These other terms, though small, are not negligible, and arise because of the slightly flattened pumpkin shape of the earth. The first of these terms is about a thousandth, and the second about a hundred-thousandth as big as the inverse-square term, at the surface of the earth. Because of the flattening of the earth's shape, the strength of the force depends to some extent on the latitude of the satellite, which is the reason for the quantities P_2 and P_4—the so-called Legendre Polynomials.

The equation given above is actually only an approximation to a far more complicated formula that takes into account a number of minor bumps and depressions in the shape of the earth. The relatively simple formula given here only describes a symmetrical spheroid. Of course, we realize that when Newton claimed that the Law of Gravitation was a simple inverse-square law, he was thinking of the force between two perfect spheres. For the kind of tests he could make this was not a bad hypothesis. After all, the closest satellite he could use for observation was the moon, for which the ratio $R/r = 1.67 \times 10^{-2}$. At the position of the moon, then, the deviation from the perfect inverse-square law caused by the oblate shape of the earth amounts to about three parts in ten million.

The Gravitational Force

Of course, the observations that Newton relied upon to test his law of gravitation were not accurate enough to detect the difference. It doesn't matter. Even if the data available had errors as great as one percent, Newton still would have said that the gravitational force is inversely proportional to the 2.000th power of the distance—not to the 1.99th or the 2.1th power—for it seemed more logical to have a simple power than a complicated one. Furthermore, it made a good analogy with the inverse-square law of the intensity of light radiating from a point source, and of course it made a perfect fit with Kepler's Third Law of Planetary Motion (regardless of the fact that the observations on which Kepler based his law were undoubtedly less than perfect).

This reasoning is an example of the fact that scientists do not develop laws of nature out of raw data alone, but instead they smooth the data over with "intuitive" ideas such as those of simplicity and symmetry. A law of nature is better if it is simple than if it is complicated. (Every student can tell when he is doing a problem incorrectly, because somehow the answer is coming out more complicated than it ought to be.)

As astronomers made more and more accurate measurements of the planetary orbits in the years following Newton's Law of Gravitation, they found that every feature of these orbits could be explained by assuming that the gravitational field of the sun was an inverse-square field. Any deviation of the planetary orbits from perfect, closed, stationary ellipses could be explained by assuming that the planets all attracted each other in pairs by means of inverse-square forces, and these attractions produced perturbations in the main motion controlled by the sun's field.

Up to the nineteenth century, Newton's Law of Gravitation was triumphant. Perfect agreement between theory and experiment to within one part out of a million proved that simplicity and symmetry were successful guiding principles in theory building.

Except for one fly in the ointment: There was something

Discovering the Natural Laws

wrong with the orbit of Mercury. It did not obey Newton's Law of Gravitation. By the nineteenth century, astronomical measurements had become so precise and the calculations of orbits had become so detailed that the effect of the planets (all the way out to Neptune) on the orbit of Mercury could be accounted for. Mercury has the most elliptical—noncircular—orbit of all the planets, and it is known from observations that the perihelion of Mercury's orbit moves around the sun at a rate of 574.10 ± .41 seconds of arc *per century*.[6] (A second of arc is 1/3600 degree.) It is also known that if you calculate the result of the perturbations due to the other planets in the solar system, there should be a precession of the perihelion amounting to 531.509 seconds of arc per century. (This includes a very small precession of 0.010 second per century caused by the oblateness of the sun, as calculated from the observed rotation of the sun.)

You can see from these numbers that there remains a motion of Mercury's perihelion amounting to 42.56 ± 0.5 seconds of arc per century that is unaccounted for by gravitational forces from any known object in the solar system. While this is an extremely small motion to worry about, it is greater than any possible error in measurement, and if our laws of nature are to be considered exact, then there is no margin for error in the theoretical calculations.

Halfway through the nineteenth century, the French astronomer Urbain Leverrier (famous for predicting the discovery of the planet Neptune by analyzing the perturbations of Uranus' orbit) attempted to explain this discrepancy by postulating an undiscovered planet named Vulcan, which was supposed to traverse an orbit between Mercury and the sun. However, Vulcan was never observed optically, and this hypothesis died a natural death. As the nineteenth century drew to a close, all attempts to explain the unaccounted-for forty-three seconds of arc as a result of observational error or perturbations from invisible sources were admitted to be failures.

The Gravitational Force

The mystery remained impenetrable until Albert Einstein, in rethinking the entire problem of gravitation, came to the conclusion that the solution lay in the law of gravitation itself—that Newton's formula, while almost perfect, was not completely perfect and not completely true. After all, the strange behavior of Mercury could be explained—as Eddington points out[7]—by replacing the exponent 2 in the inverse-square law by the number 2.00000016. But that would be too simple. Einstein recognized that in order to make a more accurate law than Newton's, it was necessary to take into account a number of effects that Newton had not put into the theory. One such effect is related to the fact that when two objects are moving so that the distance between them is changing, the force between them cannot really change until the change in the interaction has time to travel the distance between the two bodies at the speed of light. The situation is analagous to the electromagnetic interaction between two charged particles. The simple electrostatic force (the Coulomb force) is an inverse-square law but, if the charges are moving and accelerating, the force changes, and we say that magnetic and radiation forces make their presence felt.

In addition, the gravitational field has one property that the electromagnetic field does not have—a property that makes it inherently a much more complicated kind of field. This complication arises from the fact that the energy of the field itself possesses a certain amount of mass, and this mass in turn is a source of a small amount of additional gravitational field. In other words, some of the gravitational field of the sun originates in the space surrounding the sun.

Actually, Einstein, in deriving his gravitational field equations, ended up by effectively eliminating the word "force" entirely.[8] Since all objects, regardless of mass and composition, fall with the same amount of acceleration, Einstein conceived of a theory in which the *acceleration* is the important thing, and not the force. According to this theory, the path of the falling or orbiting body is determined by the curvature properties of the space through which the body is moving, and this curvature in turn is caused by the sun or a planet some distance away.

In spite of the fact that this theory fundamentally does away with concepts of "force" and "attraction," the physical result is *as if* the sun attracts the planets with a force that can be calculated by multiplying the acceleration by the planet's mass. This force can be written down as an equation similar to that of Newton; it is the famous Schwartzschild solution of Einstein's field equations, describing the field around a spherically symmetrical gravitational source:[9]

$$F = -\frac{GMm}{r^2}\left[1 + \frac{3L^2}{(mcr)^2}\right]$$

As before, G is the universal gravitational constant, M is the sun's mass, m is the planet's mass, c is the speed of light, r is the distance from the sun to the planet, and $L = mrv$, where v is the velocity of the planet as it moves in its orbit around the sun. We notice that L is none other than the angular momentum of the planet. Just as in the case of the field around the flattened earth, the first term in this expression is exactly Newton's Law of Gravitation. The second term, that varies as the inverse fourth power of the distance, is seen to depend on the velocity of the planet as well as its radius. This term is the correction to Newton's law required by the complications we have just described. It is interesting to notice that, since the kinetic energy of the planet is given by K.E. $= \frac{1}{2}mv^2 = \frac{1}{2}L^2/mr^2$, the correction term is simply $6\text{K.E.}/mc^2$.

In other words, the deviation of Einstein's formula from Newton's is just six times the ratio of the kinetic energy of the planet to the quantity mc^2—the energy of its rest-mass. Since the mass energy is very much greater than the kinetic energy, we see that this correction term is going to be rather small. We can get a better idea of how small it is by noticing that the correction term is just $3mv^2/mc^2 = 3(v/c)^2$—that is, it depends only on the ratio of the planet's speed to the speed of light.

The Gravitational Force

Let's see what this ratio is for the planet Mercury, as well as for the earth's orbit. We can show this by the following table:

	Velocity of the planet (km/sec)	Speed of light (km/sec)	v/c	$3(v/c)^2$
Mercury	48	3×10^5	1.6×10^{-4}	7.7×10^{-8}
Earth	29.8	3×10^5	9.95×10^{-5}	2.97×10^{-8}

In other words, for the orbit of Mercury the deviation of Einstein's theory of gravitation from Newton's theory is a little less than one part out of ten million.

Now how closely do the observations agree with the theoretical predictions? As we have already shown, the astronomical observations show a perihelion motion of 42.56 ± 0.5 seconds of arc that is not accounted for by the Newtonian theory. Einstein's theory predicts that there should be a precession of 43.03 ± 0.03 seconds. This agreement appears to be a remarkable triumph of both Newton's Law of Gravitation and Einstein's General Theory of Relativity. Newton's law is seen to be accurate out to at least seven significant figures, while Einstein's theory—involving that fantastically tiny correction term—is verified to about 1 percent.

However, the matter is not yet settled. During the past few years it has become apparent that Einstein's gravitational theory is not the only possible gravitational theory. An alternative theory put forth by R. H. Dicke and C. Brans[10] meets all the requirements of a proper theory, but is based upon a somewhat different set of concepts. The Brans-Dicke theory is similar to classical electromagnetic theory in that it treats the gravitational field as a traditional field of force, rather than in terms of a curvature of space-time as does Einstein's theory of gravitation. The Brans-Dicke field is much more complicated than the electromagnetic field (it is a scalar-tensor field, while the electromagnetic field is a vector field) because it must account for the effect of mass and inertia. The predictions it makes about gravitation are in general very similar to those of Einstein's theory, differing only by a few percent. You might wonder why we should pay attention to the new theory, because Einstein's theory gives almost exact agreement with the observations. The reason is that perhaps the agreement is not exact enough. In calculating the precession of Mercury's perihelion, it was assumed that the flattening of the sun's shape could be computed from the observed rotation of the sun's surface. However, a recent measurement of the sun's actual shape, made by Dicke, indicates that the sun is actually more oblate than had been expected. This greater amount of flattening gives a precession of 3.4 seconds of arc per century all by itself, causing a 10 percent discrepancy between Einstein's theory and the observations, and making better agreement between the new theory of Dicke and Brans.

However, there is still some controversy about this result, since some scientists object that the new results require a speed of rotation of the sun that is hard to believe. This is one of those subjects that is still in a state of rapid development, and undoubtedly the final answer will not be in for some time.

CHAPTER 5

The Conservation Laws: Historical Background

1. *Collisions and Momentum*

Throughout the history of physics there are intertwined two great themes: constancy and change. Even though the language might become sophisticated, so that we speak of variance and invariance, stability and instability, the basic meaning remains the same: On one hand we look for situations in which things stay the same, and on the other hand we investigate the manner in which things change. Actually, we must not think of constancy as the *opposite* of change. It is, instead, a special limiting case of change. If you have a situation in which things change so slowly or to such a small degree that the change cannot be observed, then for all practical purposes there is no change.

Newton's Second Law of Motion deals with the way the motion of an object changes as it is acted on by a force. The First Law deals with constancy and describes what happens when the applied force is reduced to zero: The velocity of the object then remains constant throughout all time. Newton's Third Law also deals with constancy. As we have already seen in Chapter 2, the Third Law is a direct consequence of a deeper law of nature—conservation of momentum—which tells us that whenever two objects interact with each other the total momentum of these two objects remains constant, as long as they are not acted on by an outside force.

As far as we can tell, this is one of the most fundamental and universal of laws. Whether it is the earth going around the sun, an electron whirling around a nucleus, two atoms colliding with each other, or two billiard balls colliding, the total momentum of these two objects remains constant, regardless of how they move or what they do to each other. Furthermore, the law applies to any number of objects interacting with each other: all the planets in the solar system, all the particles within an atom, all the molecules within a container of gas. If you multiply the mass of each object by its velocity to find its momentum, and add up all these quantities, taking the direction into account properly, you obtain the total momentum of the system—a quantity that is invariant—that does not change with time. Such an invariant quantity is said to be *conserved,* and the law that describes the conditions under which this quantity is conserved is a conservation law.

Momentum is but one of many quantities in nature that obey conservation laws. Others are angular momentum, energy, mass, and electric charge. As scientists watch the changing world about them, they maintain an eternal search to discover those rare quantities that remain constant in the midst of the change.

The belief that there are things that never change originates from an attitude that has more religion than science in it. At an early date (1644) René Descartes, the founder of modern philosophy, expressed his views:

> God in His omnipotence has created matter together with the motion and the rest of its parts, and with His day-to-day acts He keeps as much motion and rest in the Universe now as He put there when He created it . . .[1]

Within this statement there is the germ of the idea that in the universe there is some kind of quantity of motion that is conserved. However, in the early days of physics there was great confusion as to exactly how this conserved quantity of motion should be defined. Descartes chose the product of the mass and the velocity of the moving object, and it is this product mv that we now call momentum. According to Descartes' idea, if two objects (with masses m_1 and m_2 and ve-

The Conservation Laws: Historical Background

locities v_1 and v_2) collide, the quantity $m_1v_1 + m_2v_2$ is the same before the collision as it is after the collision, even though the velocities of the individual objects may have changed.

A common example of this principle is the collision of two billiard balls. If one ball makes a head-on collision with another identical ball sitting at rest on the table the first ball stops dead, while the target ball moves onward with the velocity of the original moving ball. The momentum of one ball has been transferred to the other.

Such problems concerning colliding balls were a very popular topic for speculation and experimentation during the seventeenth century, when the science of mechanics was just beginning to be understood. (We may conjecture that this interest was aroused through games such as billiards and bowling, but this would be just a speculation.) Descartes himself wrote down, as part of his laws of motion, seven rules governing the collisions between objects of various masses and velocities.[2] All but one of these rules were later proved incorrect by a little group of scientists at the Royal Society of London.

The Royal Society had been founded in 1663 to encourage experimentation in the natural sciences. One of the first activities of this group was to invite men such as the English mathematician John Wallis, the Dutch physicist and astronomer Christian Huygens, and the great English architect Sir Christopher Wren to submit papers analyzing and describing the laws of motion of colliding objects. In November of 1668 John Wallis submitted a memoir that is perhaps the first correct statement of the Law of Conservation of Momentum.

His statement was as follows: If m_1 and m_2 are the masses of two balls colliding head-on, v_1 and v_2 are their velocities before the collision, and V_1 and V_2 are their velocities after the collision, then it is an experimental fact that

$$m_1v_1 + m_2v_2 = m_1V_1 + m_2V_2$$

Now this equation appears to be very similar to Descartes' rule, but there is one very important difference. Wallis recognized that the direction of motion must be taken into account when using this equation. That is, if object 1 is moving to the right and we let v_1 be a positive number, then if object 2 is moving toward the left we must make v_2 a negative number. In modern language, we say that the velocity is a vector quantity, so that the direction of motion, as well as the speed, enters into the equation. Descartes failed to understand this distinction clearly—he used positive numbers for both velocities—with the result that his laws of impact were almost entirely incorrect.

The deepest inquiry into the problem of colliding objects was made by Christian Huygens, who worked out a complete set of rules describing the impact of objects under all conditions, including objects with unequal mass. Today we would consider all of these rules simply special cases of one general problem that we could solve by a simple application of conservation of momentum and conservation of energy. But at the time of Huygens these general laws were just being established by observing the results of specific kinds of collisions. The great generalizations were yet to be made.

A most important contribution of Huygens was his use of the principle of relativity, a scheme that is universally used today by scientists when calculating the paths of particles taking part in nuclear collisions or reactions. The basis of the scheme is this: Whenever two objects come together and collide, you can always find one particular frame of reference that moves along with the center of mass of the two objects, wherever it happens to be going. In this reference frame the center of mass of the two objects is at rest (Figure 5-1), and the total momentum of the two objects is zero, both before and after the collision. In this center-of-mass (CM) reference frame it is a relatively simple matter to calculate what the velocities of the two objects will be after the collision, if you know how fast they were going before the collision and if you also know what kind of collision took place—that is,

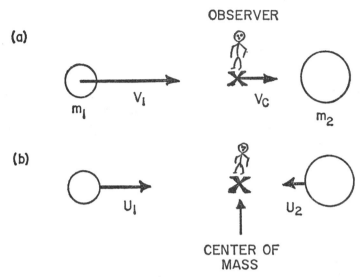

Figure 5-1. Here (a) shows two masses, m_1 and m_2, as observed in the laboratory frame of reference. Mass m_2 is at rest, while m_1 is moving toward the right with velocity v_1, relative to the lab frame. The center of mass of the system, marked with a cross, moves toward the right with velocity $v_c = m_1 v_1 / (m_1 + m_2)$. Then (b) shows the same situation as seen by an observer moving along with the center of mass. This observer is in a center-of-mass reference frame. He sees the center of mass at rest, but the two masses are moving toward each other with velocities u_1 and u_2 so that $m_1 u_1 = -m_2 u_2$—that is, the two masses have the same amount of momentum, aimed in opposite directions.

whether it was elastic, inelastic, or something in between.[3] Having found the velocities in the CM reference frame, it is a simple matter to calculate how fast the two objects are going in the frame of reference fixed to the ground (called the laboratory frame).

In the course of this work Huygens made an observation that had extremely important ramifications. He noted that when two perfectly elastic objects collided with each other, both the quantity $m_1 v_1 + m_2 v_2$ and the quantity $m_1 v_1^2 +$

$m_2v_2^2$ were the same before and after the collision. The fact that these two relationships represent two different conservation laws was not fully appreciated for many years.

Indeed, a controversy over the subject arose and persisted for over three decades. This argument was instigated in 1686 by the great German mathematician and philosopher Gottfried Wilhelm Leibniz (1646–1716),[4] who insisted that Descartes' law of conservation of motion ($m_1v_1 + m_2v_2 =$ constant) was incorrect, and claimed instead that the proper law was the conservation of *vis viva*—or "living force"—the name used for the quantity mv^2.

Nowadays we recognize that both Descartes and Leibniz were partly right and partly wrong, while Huygens was mostly right. The quantity mv is now called the momentum, and the statement $m_1v_1 + m_2v_2 =$ constant is correct only if the direction of motion is taken into account properly. As for the *vis viva*, we now recognize $\frac{1}{2} mv^2$ to be the kinetic energy of a moving object, and so in saying that *vis viva* was conserved Leibniz unwittingly gave us the first expression of conservation of energy.

The thirty-year controversy started by Leibniz raged over the question: Which is the fundamental conservation law—momentum or *vis viva?* From our modern point of view there is no controversy. Both momentum *and* kinetic energy are conserved at the same time, and each quantity has its own individual conservation law. (This is true only for perfectly elastic collisions—for inelastic collisions momentum is still conserved, but kinetic energy is not, for some of the energy goes into heat.)

I have gone into the details of this history to show that it is not an easy matter to determine the fundamental conservation laws. Furthermore, the Leibniz-Descartes controversy is an example of the kind of problem still being encountered by those studying the physics of elementary particles. Scientists search through the multitude of reactions encountered by matter and energy, and they seek those quantities that are unchanged in the midst of all the changes taking

The Conservation Laws: Historical Background

place. As we have seen, it was not immediately obvious that the quantities mv and mv^2 were individual properties of moving objects obeying separate conservation laws. Once it was fully appreciated that these were conserved quantities, it was seen that these quantities deserved names of their own, and they acquired the separate identities now called momentum and kinetic energy. (The ½ put in front of mv^2 in the formula for kinetic energy was found to be necessary in order to make the kinetic energy equal to the work done in setting the object into motion.) Ironically, now that we have reached a new level of understanding in the twentieth century, we find that the circle of opinion has gone through a full revolution and come back to the starting point. We find that Einstein's Theory of Relativity gives us a more fundamental understanding of motion and the relationship between mass and energy, and we recognize that energy and momentum are not really two separate properties of matter but may be looked on as only two components of a single four-dimensional entity.[5] From this point of view conservation of energy and of momentum are inseparable laws, and any experiment that deals with one will also have a bearing on the other.

For a discussion of this modern viewpoint we must wait until the next chapter. At this moment let us return to the original investigations of the conservation laws and see if we can find out just how accurately these laws were known to men such as Huygens and Newton. As is known by anyone who has tried to do experiments involving the timing of rolling balls or cars on a track, it is quite difficult to obtain an error of less than 1 percent without elaborate equipment. Sir Isaac Newton avoided the need to measure the speed of the colliding balls by using the pendulum experiment shown in Figure 5-2. In this experiment it is only necessary to measure the height of the two pendulums at the beginning and at the end of their swings.

Suppose that the two pendulums have equal masses. If you pull them apart, lift them to equal heights, and then let

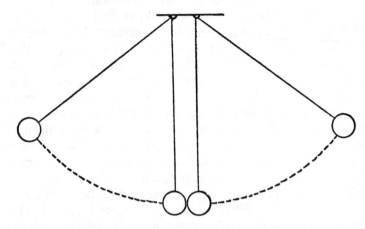

Figure 5-2. Newton's apparatus for verifying conservation of momentum. Two pendulums are arranged side-by-side, each one consisting of a ball hung by a string. The two balls are raised to the same height and then are allowed to swing together. When they rebound, they rise up to the same height, showing that they bounced apart with equal amounts of momentum.

them fall together, they will rebound. Due to air friction and imperfect elasticity they may not come all the way up to their original height, but if they are balanced to begin with, and if momentum is conserved in the collision, they will continue to be balanced after they bounce, and their heights at the end of the swing will be equal.

On the other hand, if momentum is not conserved, the effect will be as though there is an invisible force pulling these pendulums to one side or the other. One of the pendulums will rebound a bit faster than the other and so will end up at a greater height. In other words, lack of conservation of momentum would be measured by the difference in the final heights of the two balls.

Newton, always aware of the sources of error in an experiment, not only describes how to correct for air friction, but informs us that ". . . trying the thing with pendulums of ten

The Conservation Laws: Historical Background

feet, in unequal as well as equal bodies, and making the bodies to concur after a descent through large spaces, as of eight, twelve, or sixteen feet, I always found, without an error of three inches . . . that the action and reaction were always equal."[6] What this conclusion means, in modern language, is that Newton's experiment was accurate to within 1 percent. That is, a change in the system's momentum by 1 percent would have produced a three-inch difference in the height of the pendulums.

With a possible error of 1 percent in the experiment, Newton guessed at the perfect validity of conservation of momentum, and to this day he has not been proven wrong. Undoubtedly an intuitive feeling for symmetry guided Newton to his conclusion. You can see that if the universe is perfectly symmetrical, so that there is nothing pulling you more in one direction than in another, then a test particle out in empty space should stay put without changing its state of motion. If its momentum changed with time it would start accelerating away from its original position without apparent cause.

You might argue that this could happen if we were situated toward one edge of the galaxy, so that there were more stars exerting a gravitational force in one direction than in the other. Then an object left to itself would accelerate toward the center of the galaxy. But of course, all the things around this test object—including the observer—would also be accelerating at the same rate; all would be falling toward the center of the galaxy together, so that the observer could detect no relative motion. Since the force of gravitation acts on all objects with equal effect, we cannot expect any unbalance in the gravitational field of the universe to produce an observable effect on the momentum of any object left free to move of its own accord. To produce a breakdown of conservation of momentum there must be some sort of unidentified and subtle force that acts differently on the test body than on the observer, so that the result is acceleration of the test body relative to the observer.

2. Conservation of Energy

While Huygens understood conservation of kinetic energy three hundred years ago—at least as far as the collision between elastic billiard balls was concerned—it took another two of those three hundred years for the conservation concept to be extended to all kinds of energy. The eighteenth century saw a great consolidation of the energy idea, and in 1777 the connection between force and potential energy was developed by the French mathematician Joseph-Louis Lagrange (1736–1813). As is well known to students of mechanics, in any system involving objects moving under the influence of gravitational or electromagnetic forces, if you can write an equation that describes the potential energy of the system for every position of each of these objects, the rate of change of this potential energy as one of the objects moves from one place to another exactly equals the amount of force acting on the object.

For example, the strength of the gravitational force acting on an object near the surface of the earth is simply a constant:

$$F \text{ (Newtons)} = m \text{ (kilograms)} \times g \text{ (m/sec}^2\text{)}$$

where m is the object's mass and g is the gravitational acceleration (9.8 meters/sec^2). (We are making the simplifying assumption that the surface of the earth is a flat plane, so that the force does not vary with altitude; this assumption is suitable as long as we do not go too far away from the surface.) Furthermore, this force always points straight down, and in addition it does not depend on where the object is located or how fast it is moving. As a result of this, the potential energy of an object in this situation depends only on its height (h) above the surface of the ground:[7]

$$\text{Potential energy} = mgh$$

This means that it requires an amount of work equal to the quantity mgh in order to raise the mass m up to the height h,

The Conservation Laws: Historical Background

and we may imagine this work as storing energy in the gravitational field between the earth and the raised object.

The important feature of the potential energy concept is that the amount of potential energy does not depend in any way on the path taken by the object to reach its position at the time of interest. That is, a kilogram mass located ten meters above the ground has the same amount of potential energy when it is lifted straight up as it has when it is pulled up an inclined path, or when it is shot up in some sort of curved trajectory. Furthermore, if the mass is moved away from that position to some other place and then brought back to the first position, it will have exactly the same amount of potential energy as it had originally. In other words, the potential energy of an object at a given instant of time is found to depend only on its position at that particular instant, and does not depend on its past history.

Now this behavior could not be predicted theoretically—it must be established by experiment and observation. However, once it is established experimentally that the potential energy of an object does not depend on the path followed to reach its location in space, it can then be proven with complete mathematical rigor that an object under the influence of such a field of force always moves so that its *total mechanical energy* (the kinetic energy plus the potential energy) is always a constant.[8] A field of force that has this property of conserving energy is called a *conservative field*.

The gravitational field and constant electric fields are found to be conservative fields. What kind of forces are not conservative? Think of a roller coaster starting at rest at the top of its first descent, just barely making its way up to the next crest (Figure 5-3). If the second hill is the same height as the first, the car is able to return to its original altitude, where it has the same amount of potential energy. Since its kinetic energy is zero at the top of its path, its total energy is unchanged, and we see that in this fictitious situation energy is conserved.

However, in an actual situation, the second hill cannot be

Discovering the Natural Laws

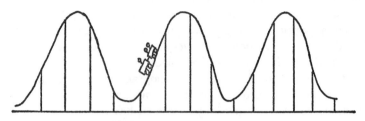

FRICTIONLESS ROLLER-COASTER

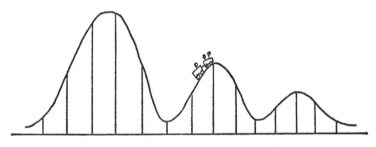

REAL ROLLER-COASTER

Figure 5-3. On a frictionless roller coaster the total mechanical energy—kinetic plus potential energy—is constant, and so the car can go back up to the same height on each hill. However, on a real roller coaster, mechanical energy is converted into heat constantly, so that the total energy diminishes as the car rolls along.

made quite as high as the first, for the car will not then reach the top. The car can only get over a lower hill, and at the top of the second hill the car has less total energy than it did when at the same height on the first hill. Where has the lost energy gone? We say that the energy of the car is reduced because of the frictional force in the wheel bearings. Friction is thus a nonconservative force.

The relationship between friction and heat, and the concept that heat is a form of energy were not recognized until

The Conservation Laws: Historical Background

the closing decades of the eighteenth century. By 1780, Lavoisier and Laplace were saying that ". . . heat is the *vis viva* resulting from the insensible movements of the molecules of a body."[9] However, it was not until 1798 that Benjamin Thompson (Count Rumford) showed by experiment that friction produced an amount of heat that depended on how much work was done in producing the friction. It thus became clear that energy was not really lost in the friction process, but was merely converted into heat energy. Once in the form of heat, it was not readily reconverted into mechanical motion, which is why it appeared to be "lost."

Yet it required another half-century before conservation of energy was to be established as a general law. Perhaps one reason for the long time lag was the fact that many scientists did not really believe in the concept of molecules except as a convenient hypothesis to explain certain observations. As Max Jammer points out,[10] the Royal Society of London managed, in 1845, to perform one of the classical mistakes of scientific politics by rejecting a paper of J. J. Waterston, the first calculation of the average energy of a system of molecules colliding at random with one another, thus setting back the development of molecular physics by at least a decade. The hard-headed conservatives of the scientific world were reluctant to deal with such an unobservable thing as a molecule. In part, this attitude was an overreaction to the fanciful theory spinning of pre-Newtonian centuries.

For such reasons, it was necessary for the concept of conservation of energy to be built up laboriously by a long series of experiments involving the energy going into and coming out of all the different kinds of reactions known. Not only the energy of friction was involved, but the energy going into electrochemical reactions, as well as the conversion of electrical energy into heat. One of the motivations contributing to these experiments was a philosophical movement known as *Naturphilosophie* that was popular in Germany. As early as 1799, Professor Friedrich Wilhelm Joseph von Schelling predicted that ". . . magnetic, electrical, chemical, and

Discovering the Natural Laws

finally even organic phenomena would be interwoven into one great association . . . [that] extends over the whole of nature."[11] Among Schelling's students were the scientists whose energy-balance measurements culminated in the final statement of the Law of Conservation of Energy. The key measurement was the demonstration, by numerous experiments, that to every quantity of heat there corresponds a definite amount of mechanical work, or energy.

Eric Rogers[12] provides a long list of experiments devoted to measuring this mechanical equivalent of heat, starting with Count Rumford's crude cannon-boring feat, progressing through Joule's exhaustive measurements during the 1840s, and culminating with a determination by a group at the U. S. National Bureau of Standards in 1939. This last experiment was actually a measurement of the electrical heating of water, so that in this case we have a situation where the mechanical work is done on charges (electrons) being accelerated in an electric field, and the number of calories of heat produced is measured in the usual way by the rise of temperature of a known amount of water. A value of 4.1819 joules per calorie was obtained, as compared with the value of 4.1802 obtained in 1927 by measuring the rise in temperature of water churned with a paddle wheel. Thus two different sources of mechanical and electrical energy gave amounts of heat that agreed to within 0.05 percent.

Of course, if we insist on dealing only with visible, surface phenomena, we can never fully prove the Law of Conservation of Energy—because how do we know that over the horizon there might not be discovered a new kind of reaction that behaves altogether differently from the old reactions already tested?

The problem is narrowed down to a much more restricted question if we take the modern point of view that all matter consists of elementary particles, and that all the visible actions and reactions that matter enters into are the end results of the interactions going on between the individual elementary particles. Now if we can show that all of these particles in-

The Conservation Laws: Historical Background

teract by means of conservative forces—that is, if there is no energy gained or lost during each of these elementary interactions—then there can be no energy gained or lost on a macroscopic or visible level.

The only thing that remains to be shown is that all of the fundamental forces that exist in nature are indeed of the conservative type. This problem divides itself into two parts: first, to show that the four known interactions conserve energy; and second, to determine that the four known interactions are the only interactions (or else, if it turns out that there are new interactions to be found, to show that there are still no hidden means for energy to be created or destroyed). The study of energy conservation thus descends to the nuclear and subnuclear level, and in the next chapter we will see that some of the latest evidence shows energy to be conserved to an extremely high degree of accuracy, at least for the known interactions.

As for new and as yet unknown interactions, this is a more difficult question to answer. It was, in fact, this very line of reasoning that led Fred Hoyle to postulate the existence of a kind of force field that had the property of creating particles of matter in order to form the basis of his steady-state theory of the universe.[13] Since the creation of a particle requires the creation of an equivalent amount of energy, the steady-state theory was forced to accept a small but finite breakdown of the energy conservation concept. Hoyle's theory required the creation of one atom of hydrogen per second in a cube of space 160 kilometers on a side. The question is, how much of a breakdown in the law was required?

Now one atom of hydrogen per second in a cube with a volume of 4.1×10^{15} cubic meters seems like a very small quantity of matter to pop up here and there. However, a different perspective on the problem is obtained if we think about the total rate at which energy is created in the stars in our immediate neighborhood, within one hundred light-years of the earth. We find that the rate of energy production, averaged over the entire volume of space occupied by those stars, is 1.5×10^{-24} watt per cubic meter.[14]

We can use the Einstein relation $E = mc^2$ to find the amount of mass represented by this energy production, and obtain a rather surprising result. In spite of the fact that the stars burn up vast amounts of matter each second, the volume of space is so enormous that the energy produced per second has a mass density of only about 1.7×10^{-41} kilogram per cubic meter. This figure represents the mass of about forty hydrogen atoms converted to radiant energy each second in a cube 160 kilometers on a side (compared to the one per second required by Hoyle). Looking at it another way: Hoyle's theory of the expanding universe required the spontaneous creation of matter-energy to take place at about 2 percent of the rate at which energy is produced by thermonuclear reactions in the stars.

Such a violation of conservation of energy is not trivial. You cannot sweep it under the rug by saying that our measurements of energy conservation are too inaccurate to rule out such a small violation. Instead, it is necessary to assert boldly, as Hoyle did, that this theory requires the hypothesis of a new kind of force that simply does not obey the Law of Conservation of Energy.

To test such a hypothesis requires us to test the steady-state theory of the universe, together with the alternate big-bang theory and oscillating universe theory. The main type of data used for distinguishing between the theories are counts of the number of galaxies as a function of distance away from the earth—or, to be more precise, the density of galaxies throughout the universe. The steady-state theory predicts that the density of matter in the universe is uniform throughout. There are also measurements of electromagnetic radiation in the centimeter-wavelength range floating around through space. Such radiation is expected to be related to the release of energy during the primordial explosion of the big-bang theory.[15]

Interpretation of the data appears to depend somewhat on

The Conservation Laws: Historical Background

the political position of the interpreter. The consensus of opinion appears to be that the steady-state hypothesis has not been upheld, but there are still arguments raised by the steady-state supporters. The final answer undoubtedly will come after many decades of arduous data-taking and analysis have pinned the numbers down to the point where all the hypotheses except one have been demolished.

3. *Conservation Laws and Symmetry*

One of the more elegant results of the science of mechanics is the demonstration that for every conservation law there exists a particular kind of symmetry of space or time.[16] This connection between conservation and symmetry arises because of the fact that the total energy of a system of objects can be written as the sum of the kinetic energy and the potential energy. The potential energy, in turn, depends on the kinds of forces acting on these objects—whether they arise from a gravitational field, electromagnetic field, or other types of fields. If you can write down an equation showing how the potential energy depends on the position and velocities of the various objects in the system, then—in principle, at least—the future motion of these objects can be predicted.

For example, if one or more objects are moving about in a gravitational field close to the surface of the earth (close enough so that we can consider the surface a flat plane), then the potential energy depends only on the height of each object above the ground. It does not depend on where the object is located parallel to the ground—that is, in a north-south-east-west direction. If this is the case, it can be shown with complete rigor that the total momentum of these objects in a direction parallel to the surface of the earth must always remain constant. The momentum perpendicular to the surface is not constant, for the force of gravity acts in the downward direction, but since the potential energy does not depend on the horizontal location, there is no force acting in the horizontal direction.

Furthermore, if the potential energy does not depend on

Discovering the Natural Laws

the location of these objects in the horizontal plane, it makes no difference where the origin of the reference frame is located. In other words, the space we are considering is symmetrical in the horizontal plane. You see that here is a symmetry condition that is closely connected to the Law of Conservation of Momentum.

If we go out into intergalactic space and postulate that space is symmetrical in all directions and that the potential energy is constant no matter where we are located, it follows that momentum is conserved in all directions—there is no tendency for an object to move in one direction or another, and the center of mass of a system will always remain in one place (or at least will move with constant velocity).

Another type of symmetry deals with time. If the potential energy of a system does not depend on time in any way, so that it does not matter where we start the zero of our time scale, the result is that the total energy of the system is constant. In other words, symmetry with respect to time results in conservation of energy.

Now there is a great temptation to say that it stands to reason—it is "intuitively obvious"—that space and time are symmetrical, and therefore the conservation laws must be perfectly true. However, such arguments ignore the order in which our knowledge is acquired. First we learn about conservation laws by making observations on material objects. From these conservation laws we decide that nature has certain symmetries. To turn the argument around and invoke nature's "love" of symmetry is as bad as proclaiming that nature abhors a vacuum to explain why a pump fails to work.

Only people love. Nature just is.

CHAPTER 6

The Conservation Laws: Modern Experiments

1. *Excited States of Atoms*

Imagine an elevator being raised to the top floor of a tall building. Work is done in lifting the elevator against the pull of gravity, and we say that the elevator has been raised to a region of greater potential energy. In this way the total energy of the system (consisting of the earth and the elevator) has been increased. If the elevator is not supported by some kind of force, it immediately falls back to the ground, converting its potential energy into kinetic energy and ultimately into heat (Figure 6-1). Measurements of this kind of energy

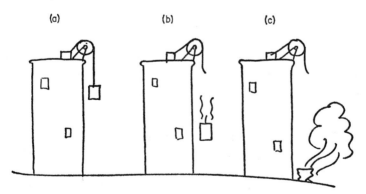

Figure 6-1. An elevator is raised to a height H, so that its potential energy is mgH. In (b) the elevator falls. While it is falling the sum of its kinetic and potential energy is always equal to mgH. In (c) the elevator has come to rest, and the quantity of energy mgH has been converted into heat.

Discovering the Natural Laws

transfer during previous centuries convinced scientists that the amount of energy released by the system in crashing back to the ground is equal to the energy that was put into the system while raising it to its higher state.

Raising a weight against the force of gravity is but one example of a type of event very common in nature. There are many kinds of situations in which a quantity of energy is absorbed by a system, which is then said to be in an *excited state*. In the course of time this energy is subsequently released from the system, so that it returns to its normal, stable *ground state*.

Raising an electron from one atomic orbit to another is such a situation. In atomic and subatomic systems there is the additional complication that the electron will only allow itself to be moved to certain very particular orbits in which it possesses very special quantities of energy. In other words, the electron can move from orbit to orbit only in quantum jumps, and the amount of energy required to make a given electron jump from its ground state to a particular orbit is a very definite amount of energy.

For example, if an electron beam is sent through a tube of hydrogen gas so that the electrons collide with the hydrogen atoms, some of the orbital electrons will be bounced up into excited states (Figure 6-2). However, this excitation of the atomic electrons cannot take place unless the electrons in the beam have an energy of at least 10.2 electron-volts. This is the energy of the lowest excited state in the hydrogen atom.

Now this excited state is an unstable state; the atom would like to return to its ground state as quickly as it can, and this it does by emitting a pulse of electromagnetic waves—a photon of light. By measuring the wavelength of the emitted photon with a spectroscope, we can determine the energy given off by the atom in returning to its ground state (see Appendix I). This energy turns out to be just 10.2 electron-volts, as closely as can be measured, in agreement with the energy used to excite the atom in the first place. You can see that if

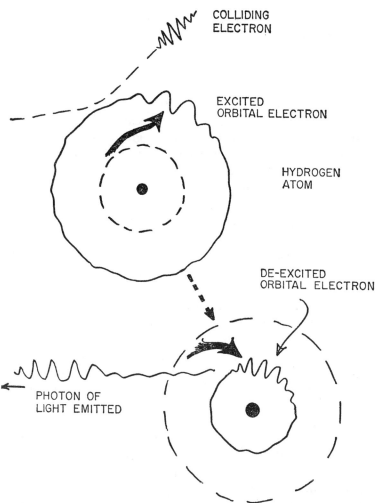

Figure 6-2. An energetic electron collides with a hydrogen atom in a container of hydrogen gas. The orbital electron is propelled up into an excited state. After a short time this orbital electron falls back to its normal ground state, getting rid of its extra energy by emitting a photon of light. In the case illustrated, this would be a photon of ultraviolet light. The electrons are shown as wave packets, to symbolize that the rules of quantum mechanics apply in this atomic world.

it takes a certain amount of energy to raise an atom to an excited state, and that later on exactly the same amount of energy is given off, this is evidence that the atomic binding force attracting the electron to its nucleus is a conservative force—one that obeys the laws of conservation of energy. In raising the elevator to the top of the building, the force involved was the gravitational force—because as the elevator was raised, energy was stored in the gravitational field and subsequently returned in the form of kinetic energy. In the case of the atomic excitation, the force involved is the electromagnetic force between the electric charges of the orbital electron and the atomic nucleus.

One trouble with this kind of measurement is that you cannot measure the energy used to excite one particular atom, and then measure the energy emitted by that same atom. The atoms have no labels, and all you do is measure the energy going into an enormous number of atoms, together with the energy coming out as photons of light.

Another complication that arises with this kind of measurement is that the atoms giving off the light are moving around in the gas at a high speed, due to their high temperature. When light is given off from a moving source its wavelength is changed. This is the well-known Doppler effect: If the source is moving toward the observer the wavelength is decreased, and vice-versa. Since all the atoms in the gas are moving about in all directions, the result is that the light reaching the spectroscope covers a rather wide band of wavelengths, even though the light emitted from each atom has a much more definite wavelength. This means that the bright line observed in the spectroscope has a width of one or two Angstrom units ($1 \text{ A} = 10^{-10}\text{m}$) instead of the much narrower natural width that might be expected. (This effect is known as *Doppler broadening* of the spectrum line.) As a result of this line broadening, the observer cannot be sure that the photon coming out of a given atom has exactly the amount of energy that was put into the atom; there could be a difference of a few tenths of an electron-volt.

In recent years a method has become available for overcoming at least part of this problem, so that only those atoms that emit photons in a very narrow energy range are detected. This method makes use of the fact that excited states do not decay immediately to the ground state, but take more or less time, depending on the energy and angular momentum of the state. A level normally decays in approximately 10^{-8} second, but it may take more than ten times longer if the emission of the photon is "not allowed." This situation makes it possible to prepare a supply of atoms suspended in their excited states for a brief period of time.

What makes this interesting is the fact that if a photon comes along with exactly the same energy as the excited state, this first photon can stimulate the emission of a second photon of light, and thus two photons arise where formerly there was one. This kind of action is the basis for the operation of the laser, the device whose name is an acronym for Light Amplification by Stimulated Emission of Radiation. The construction and operation of lasers is described in numerous books,[1] and their uses in scientific research are highly varied.

Our purpose at this moment is to point out that the very fact that lasers work, producing light with a very accurately defined wavelength, is direct evidence for conservation of energy. There are three energies that have to be matched in the operation of a laser. First of all there is the energy of the excited state, which requires a photon of exactly the right energy to come along to stimulate its de-excitation. The emitted photon has the same energy as the stimulating photon, and in turn it has exactly the right amount of energy to stimulate another atom—in an excited state with the same energy—to emit another photon.

An essential part of the laser operation is the fact that all of these activities take place in a long tube with mirrors at two ends. The mirrors cause the tube to act as a resonant cavity; if the length of the tube is a multiple of the photon wavelength, these photons will bounce back and forth, will be trapped efficiently, and will cause the emission of more photons from the excited atoms. Photons having the wrong wavelength or going in the wrong direction will be lost. In this way the resonant cavity picks out of the broad range of energies present only those that correspond to the resonant wavelength.

Even with the most careful design a laser cannot be built that will emit light with only a single wavelength and energy. There is always a small "bandwidth" or "linewidth" that results from a number of factors: the width of the atomic energy level itself, the diameter of the laser mirrors, and thermal vibrations in the laser itself causing fluctuations in the distance between the mirrors. The helium-neon gas laser can be built to have an extremely narrow linewidth: The spread in wavelength divided by the actual wavelength may be as small as 10^{-13}. This means that, to within one part out of 10^{13}, the energies of the levels and the photons involved all agree.[2] Reportedly, some lasers have achieved "monochromaticity" (that is, the ability to emit single wavelengths) to a precision better than one part in 10^{14} for a short time.[3]

These figures, then, represent the limit to which conservation of energy has been verified in reactions involving excited states of orbital electrons, in which the binding force is an electromagnetic one.

2. Excited States of Nuclei: The Mössbauer Effect

The effects described in the previous section occur not only in the outer electron orbits of atoms, but also within their nuclei. The neutrons and protons within a stable atomic nucleus (collectively known as nucleons) are held together by means of the strong nuclear force, an attraction that opposes the electrostatic repulsion of the protons. Just as the electrons in the orbital shells of the atom may exist in certain specified energy levels, the nucleons within the "potential energy

The Conservation Laws: Modern Experiments

well" of the nucleus may exist in certain particular energy levels. In the stable, or ground state of the nucleus, all of the lowest levels are filled up with nucleons, but there are higher levels into which the nucleons can be driven temporarily by the addition of the right amount of energy (Figure 6-3).

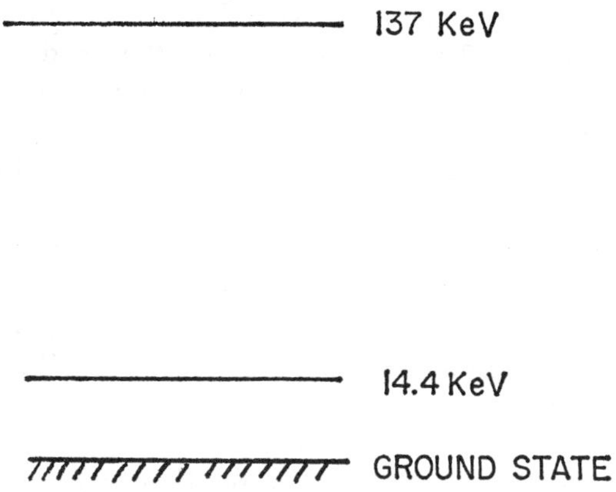

——————————————— 137 KeV

——————————————— 14.4 KeV

////////////////// GROUND STATE

Figure 6-3. A schematic diagram of the two lowest excited states in the iron-57 nucleus. The nucleus is normally in the ground state. However, if a 14.4 keV gamma ray photon comes along and interacts with the nucleus, this energy will be absorbed, and the nucleus will be excited into the 14.4 keV state. Subsequently it gets rid of this energy by emitting a 14.4 keV gamma ray and going back to the ground state.

One of the ways in which this can happen is for a photon to come along and give its energy to one of the particles within the nucleus. If this energy is not enough to drive this nucleon entirely out of the nucleus, it may be enough to lodge the particle temporarily in one of the excited states—one of the upper rungs of the energy ladder. In order for this to happen, the incoming photon must have an amount of energy that exactly matches the energy of the excited state, for this is a resonance effect. It is the same kind of effect by which a radio or television receiver picks just one frequency out of the many waves that may be passing through the antenna at any given moment, arriving from the numerous transmitters in the neighborhood.

However, as is well known, a given radio transmitter does not send out a single well-defined frequency, but instead broadcasts an entire band of frequencies necessary to convey the information being transmitted. The width of the band depends on how rapidly the signal modulating the carrier wave is varying. That is, the higher the modulating frequency, the greater the bandwidth required to transmit the signal. For this reason, television signals require a greater bandwidth than ordinary music broadcast on your local radio station.

In an entirely analagous manner, if an excited state in a nucleus decays in a very short time, there is a wide spread in the energies of the gamma rays given off by a number of identical nuclei excited to that state. On the other hand, if the excited state lasts for an appreciable length of time, then the width of the level is very narrow.

The narrow level width means two things: First of all, the gamma rays given off by the decay of nuclei from such a narrow level are very accurately defined in energy—that is, the spread of energy is very small. Second, if the experimenter tries to excite such a narrow level by the resonance absorption of gamma ray photons, only those photons within a very narrow range of energies will be absorbed by the nuclei to produce the excited state.

How do we make use of this effect? Consider the following situation: We take a small quantity of Co^{57}, a radioactive isotope of cobalt that can be made by bombarding iron with deuterons (H^2) in a cyclotron, according to the reaction $Fe^{56} + H^2 \rightarrow Co^{57} + n$. (Alternatively, the reaction $Ni^{58} + p \rightarrow Co^{57} + 2p$ may be used.) Co^{57} has a half-life of 270 days, and decays by means of a process known as electron capture. This process occurs because the Co^{57} nucleus has in it too many protons to be stable. It is able to neutralize one of these protons by capturing one of the elec-

The Conservation Laws: Modern Experiments

trons from the innermost atomic orbit surrounding the nucleus, thus forming an atom of Fe^{57}, which has the same mass as Co^{57}, but one less nuclear electric charge.

The nucleus of Fe^{57} formed in this way is not yet in its stable ground state. It still has some energy to get rid of before it can remain permanently at rest. It gets rid of this energy by emitting one or two photons, and so drops down the energy level ladder. When the orbital electrons of an excited atom go through this process, the energies involved are only a few electron-volts, and so the light emitted falls into the visible or ultraviolet part of the spectrum.

The energies involved in going from one nuclear level to another may range from a few thousand to millions of electron-volts, and this means that we are talking about the emission of gamma rays from the atomic nucleus. The nucleus of Fe^{57} resulting from the decay of Co^{57} is in an excited state about 137 keV above the ground state. It is found experimentally that in about 9 percent of the cases these nuclei drop down directly to the ground state with the emission of a 137 keV gamma ray. Most of the time two gamma rays are emitted, one after the other; the nucleus first gives off a 123 keV gamma ray, dropping to a level existing 14.4 keV above the ground state, rests there for an average of 1.4×10^{-7} second (0.14 microsecond), and then drops the rest of the way with the emission of a 14.4 keV gamma ray. It is this second gamma ray that we want to fix our attention on, for it is the key to the entire energy-measuring scheme we are discussing.

Suppose we have a Co^{57} source emitting these 14.4 keV gamma rays and we place nearby a gamma ray detector (for example, a scintillation counter) that gives off an electrical pulse every time a gamma ray enters it. These electrical pulses are sorted out in a pulse-height analyzer that is able to separate the pulses produced by the 14.4 keV gamma rays from the larger pulses resulting from the 123 keV gamma rays. Since 14.4 keV is quite a low energy as gamma rays go (in fact, its energy is less than that of the

X-rays emitted by the average television set), it is necessary to use a very thin source so that the gamma rays are not stopped within the source itself.

The next step is to put an absorber containing Fe^{57} atoms between the radioactive source and the detector. (Natural iron contains 2.19 percent Fe^{57}; iron enriched in Fe^{57} can be obtained.) Gamma rays of such energies are absorbed in matter mainly by the photoelectric effect: Each gamma photon shoots through the absorbing material until it collides with an atom. During this collision it gives all its energy to one of the orbital electrons, knocking it violently out of its orbit. In so doing, the photon disappears.

In the special situation we are discussing, there is another way for the gamma ray photons to be absorbed; this is the resonance absorption effect, in which a target nucleus absorbs a photon and is raised to one of its excited states. In order for this to happen, the gamma ray energy must exactly match the energy of the excited state. One way to do this is to make the absorber out of the same material as the source emitting the gamma rays. In the case we have been discussing, the emitter is an excited state of Fe^{57}, and so we make the absorber out of Fe^{57}, which can be raised to the 14.4 keV excited state by absorption of a 14.4 keV photon. By means of this process the absorber removes more photons from the beam of gamma rays than would be removed by the photoelectric effect alone, and this is detected by a reduced counting rate in the detector.

Although there are numerous radioactive elements that give off gamma rays of all energies, this resonance absorption effect is not noticed in most of them. The reason for this is the fact that each gamma ray photon emitted from a nucleus carries with it a certain amount of mass, energy, and momentum. The emitting nucleus (to conserve energy and momentum) must recoil, just as a gun recoils when it shoots a bullet. This means that some of the energy from the exited state goes into kinetic energy of the recoiling nucleus, and as a result the emitted photon does not have all

The Conservation Laws: Modern Experiments

of the energy that originally came from the excited state. For example, consider a hypothetical nucleus of atomic weight 100, having an excited state with an energy of 100 keV. As it gives up this energy by emitting a gamma ray, the recoil energy of the nucleus is about 0.05 electron-volt. This means that the photon energy is less than the energy of the excited state by 0.05 eV out of 100,000 eV, or five parts out of ten million.

If this excited state has a typical lifetime of about 10^{-7} second, its level width is about 5×10^{-9} eV, according to the Heisenberg uncertainty relationship (see Appendix I). That is, if this level is to be excited by resonance absorption of a photon, the photon energy must be 100 keV plus or minus 5×10^{-9} eV; its energy must equal the level energy to within five parts out of 10^{14}. Resonance absorption is an exceedingly fine energy discriminator, so fine that the 0.05 eV of energy lost by one of the above-mentioned photons as it kicks its nucleus away puts it way out of the acceptable range for absorption. As a result of the recoiling nucleus, resonance absorption of gamma rays is not detected under ordinary circumstances.

However, there are certain special conditions under which this recoil does not take place. The conditions required are: 1. that the nucleus emitting the gamma ray be bound firmly in a crystal lattice structure, and 2. that the energy of the emitted gamma ray be not too great. If these conditions are met, the emitting nucleus is not jarred out of its position in the crystal lattice but hangs in there while the recoil is taken up by the crystal as a whole. The mass of the entire crystal is so great that the recoil energy is quite negligible, and so the emitted photon has all of the energy of the excited state. Now when this photon encounters a suitable nucleus in an absorber, it has just enough energy to pop the absorbing nucleus into its excited state.

For example, sources of Co^{57} are prepared by plating the radioactive cobalt onto thin iron foils, with proper heat treatment. The Fe^{57} formed by the decay of the Co^{57} then finds itself part of an iron crystal and so is able to emit its 14.4 keV photon without recoil. A similar iron foil (enriched in Fe^{57}) can then act as absorber for the resonant photons.

It is usually necessary to make some special arrangement to tell how much of the photon attenuation is caused by the nuclear resonance effect in which we are interested, how much is caused by ordinary photoelectric effect, and how much is simply a result of scattering of photons out of the path of the detector. To do this we make use of the fact that motion of the source either toward or away from the absorber causes the frequency—and thus the energy—of the photon to be raised or lowered by means of the Doppler effect. So sensitive is the absorption in Fe^{57} to a change in photon energy that moving the source with a speed of 0.017 cm/sec reduces the absorption to one-half its maximum value. By plotting a curve of absorption vs. source speed, we are actually making a measurement of the width of the nuclear energy level (Figure 6-4).

The process of recoilless emission and resonance absorption of gamma rays is known as the Mössbauer effect, in honor of Rudolf L. Mössbauer, who first discovered and made use of this phenomenon in 1957, during his graduate work in Heidelberg.[4] The extreme narrowness of the energy levels within the nucleus makes the emitted photons monoenergetic to a degree unmatched elsewhere. These photons thus become an energy probe that can be used as a tool for measuring energies and energy changes with extreme precision. A great number of uses have been found for this tool, but in this book we are interested in the fundamental fact that the Mössbauer effect is in itself a demonstration of conservation of energy, as well as conservation of momentum.[5]

In order for the Mössbauer effect to work the way it does, several things are necessary: First of all, the millions of Fe^{57} nuclei must have identical energy levels (or at least levels that differ in energy by no more than the level width ΔE). Second, there is no loss of energy as the excited state emits a photon while dropping down to the ground state. Third,

The Conservation Laws: Modern Experiments

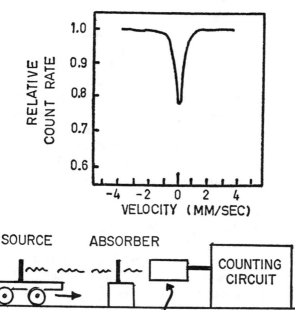

Figure 6-4 top. The data from a typical Mössbauer effect experiment and the experimental apparatus. As the gamma ray source is moved slowly toward or away from the absorber, the number of photons getting through the absorber varies. The absorption is greatest at zero velocity, so that at that point the number of counts is a minimum. The spread in gamma ray energy is measured by plotting a curve of counting rate vs. source velocity.

the photon loses no energy in transit from the emitter to the target nucleus, and finally the photon gives all its energy to raise that target nucleus to its excited state. The emission, propagation, and absorption of a photon are phenomena caused and controlled by the electromagnetic interaction. Therefore, the Mössbauer effect demonstrates conservation of energy in electromagnetic interactions. In addition, we note that the amount of energy in a nuclear excited state is determined heavily by the strength of the strong nuclear force. Since a certain quantity of energy can raise a nucleus to a given level, and then after some delay the same amount of energy is given off, this evidence tells us that the strong nuclear force is also a conservative force.

All of this argument tells us that to find how accurately the Law of Conservation of Energy is known, we must look for the narrowest nuclear energy level whose width has actually been measured. Such a measurement becomes extremely difficult when the level width is very narrow, because the resonance absorption might be wiped out by sound vibrations within the emitter or absorber, or by electric and magnetic fields from other atoms in the crystal containing the emitting nucleus.

The narrowest well-established level appears to be that of Zn^{67}, which has a 93 keV level with a lifetime of 9.4 microseconds.[6] The search for this resonance was made using a rigid package of source and absorber suspended by threads inside a liquid helium cryostat. Cooling the material to almost absolute zero reduced the Doppler effect due to the thermal motion of the atoms. The width of this level is expected to be 5×10^{-11} eV according to the Heisenberg uncertainty relationship. Early measurements were merely able to show the existence of resonance absorption, and to show that this resonance disappeared when a small magnetic field was applied, changing the energy of the level.

More recent experiments[7] have succeeded in actually measuring the line width by means of the Doppler shift technique. The measured width appears to be three to five times the theoretically expected width. This broadening depends on the method of preparing the source and can be blamed on the interaction between the nuclei and the crystal lattice in which they are suspended. In any event, these experiments determine the photon energy and thus verify conservation of energy to within a few parts out of 10^{15}.

There is actually evidence of a much narrower energy level that can be excited by nuclear resonance absorption of photons. The isotope Ag^{107} results from the decay (either by positron emission or electron capture) of the radioactive

The Conservation Laws: Modern Experiments

nucleus Cd^{107}. The Ag^{107} is first formed in an excited state having an energy of 93.5 keV. It waits in that excited state for a mean lifetime of 63 seconds before dropping down to the ground state. As a result of this very long lifetime, the theoretical level width is 1.0×10^{-17} eV. If this width could actually be measured by the Mössbauer effect, it would show conservation of energy to one part out of 10^{22}. However, magnetic fields from atoms neighboring the emitting nuclei are expected to broaden the energy level by a factor of 10^5, so that one part out of 10^{17} should be the measurable limit for the fractional width of this level.

The Ag^{107} level has been studied by a group of physicists at the USSR Academy of Science.[8] A Cd^{107} source was produced by irradiating Ag^{107} with 17 MeV protons in a cyclotron. The radioactive silver, together with a 99.999 percent pure silver absorber was suspended in liquid helium at a temperature of 4.2°K. Some of the 93.5 keV gamma rays were absorbed by nuclei in the silver absorber, which were raised to their 93.5 keV excited state. In this experiment the absorption was not measured by observing the reduction in the number of gamma rays passing through the absorber, as in the usual Mössbauer effect experiment. Instead, advantage was taken of the unusually long lifetime of the excited state. The absorber was removed from the radioactive silver after three minutes' exposure, placed next to a scintillation counter, and 93.5 keV gamma rays counted as the excited nuclei dropped down to their ground states. In this experiment, the actual line width of the excited state was not actually measured. (Plans for making this measurement were announced by Bizina *et al;* but as of the date of writing have not yet been published.) However, measurement of the cross section for exciting the nucleus agreed roughly with the expected value based on the theoretical level width.

Since perturbations due to the motions of nuclei, their innate magnetic fields, and other disturbing influences appear to limit the measurement of level widths to the order of one part out of 10^{17}, it may be that such "noise" factors present us with a natural barrier preventing confirmation of the conservation laws to a greater degree of accuracy.

3. *Nuclear Reactions*

In the Mössbauer effect, the nucleus involved in the reaction does not change in form—it merely goes from one state to another. Does conservation of energy hold equally well for reactions in which the nucleus changes from one form to another? In the average book on nuclear physics, energy and momentum are considered to be perfectly conserved, and all of the equations used in analyzing experiments are based on that assumption. However, according to the point of view we are emphasizing here, we must look for the nuclear reaction experiments in which energy and momentum have been measured most accurately, and analyze the data to see if the energy and momentum of the particles going into the reaction are the same as the energy and momentum of the particles coming out of the reaction. From then on we are justified in using the conservation laws to analyze other measurements of lesser precision.

Actually, in the Theory of Relativity, energy and momentum are both so intimately connected that if one is conserved, the other must be conserved also. (The three components of momentum together with energy make up the four components of a four-dimensional vector.) This is demonstrated by the equation relating energy to momentum:

$$E^2 = (mc^2)^2 + (pc)^2$$

where E = the *total energy* of an object or system of objects with rest mass m and momentum p (see Appendix II).

Since the rest mass is a constant, then if the total energy remains constant during a reaction, the total momentum must also remain constant. Of course, this conclusion is based on the validity of the principle of relativity—in particular the equivalence of mass and energy. (The principle of relativity will be discussed more fully in Chapter 7.) Conversely, if we measure the energy, momentum, and masses of the par-

The Conservation Laws: Modern Experiments

ticles engaged in a nuclear reaction, the most accurate of these measurements do indeed verify that the conservation laws together with the principle of relativity properly describe the events that take place.

A typical experiment is performed by means of an arrangement such as the one illustrated in Figure 6-5. A target of

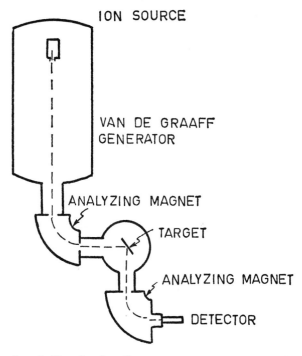

Figure 6-5. A Van de Graaff generator shoots a beam of protons or deuterons into a target. The energy of the particles hitting the target is measured by the first analyzing magnet. The energy of the particles coming out of the target is measured by the second analyzing magnet.

the element to be studied is bombarded with charged particles (such as protons or deuterons) that have been propelled to a great velocity by a device such as a Van de

Discovering the Natural Laws

Graaff generator. (A cyclotron may be used for such experiments, but Van de Graaff generators are less expensive and have more precise energy control.) The energy of the particles hitting the target is measured by bending the path of these particles through a magnetic field of known strength. This magnetic field is produced by an accurately regulated electric current passing through a coil of wire wrapped around an iron pole-piece. Only those particles that travel along a curve of fixed radius so that they pass through a narrow exit slit are allowed to strike the target.

The energy of the outgoing particles is likewise measured

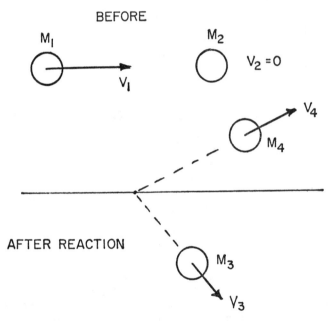

Figure 6-6. Before-and-after diagrams of a nuclear reaction. M_1 is the mass of the incident particle, while M_2 is the mass of the target particle. M_3 and M_4 are the masses of the particles produced in the reaction. The velocity vectors are as shown for one particular reaction. Of course, the products of the reaction may be emitted at any angle consistent with the conservation laws.

The Conservation Laws: Modern Experiments

by bending through another analyzing magnet. These outgoing particles may be either particles elastically scattered from the incident beam by the nuclei in the target, or else particles produced as a result of nuclear reactions occurring when incident particles actually enter the target nuclei. The analyzing magnet generally is built so that it can be swung around on a pivot so as to accept particles coming out of the target at any angle. In the experiments to be discussed we will consider only particles coming out at an angle of 90° with respect to the incident beam.

A nuclear reaction can in general be pictured as shown in Figure 6-6. We consider a particle of mass M_1 coming along with velocity v_1 and reacting with a particle of mass M_2 at rest. Coming out of the reaction are two particles with masses M_3 and M_4, having velocities v_3 and v_4. Since we are dealing with particles having great amounts of energy, we must be careful to use the proper relativistic formulas for mass and energy. The masses defined above are the total relativistic masses, which are greater than the rest-masses. For a given particle, the relativistic mass M is related to the rest-mass m by the expression

$$M = \frac{m}{\sqrt{1 - v^2/c^2}}$$

The total energy E of a particle may be divided into two parts: the rest-mass energy E_0 and the kinetic energy T. Each of these quantities of energy is related to a quantity of mass by the formula describing the equivalence of mass and energy. Thus, the total mass is related to the total energy of a particle by the expression $E = Mc^2$, while the rest-mass and the rest-mass energy for the same particle are related by the formula $E_0 = mc^2$.

In all nuclear reactions the total mass before the reaction must equal the total mass after the reaction. In the reaction we are talking about, this Law of Conservation of Mass is expressed by the equation

Discovering the Natural Laws

$$M_1 + M_2 = M_3 + M_4 \tag{1}$$

If we multiply both sides of this equation by c^2, we see that this same equation expresses the Law of Conservation of Energy; the total energy (rest-mass plus kinetic) before the reaction equals the total energy after the reaction:

$$E_1 + E_2 = E_3 + E_4 \tag{2}$$

Of course, the rest-mass energies before and after the reaction are not equal. If the particles leaving the reaction have more energy than the particles coming in (as in an exoenergetic reaction), then those outgoing particles have less rest-mass than the incoming particles. The leftover mass belongs to the energy Q released by the reaction:

$$E_{01} + E_{02} = E_{03} + E_{04} + Q \tag{3}$$

Since the total mass of each particle may be written as its rest-energy E_0 plus its kinetic energy T, we can express Q in terms of the kinetic energies of the particles taking part in the reaction:

$$Q = T_3 + T_4 - (T_1 + T_3) \tag{4}$$

(See Appendix II for details of all the derivations in this section.) The energy released is seen to equal the kinetic energy of the outgoing particles minus the kinetic energy of the incoming particles. We can also express Q in terms of the rest-masses of the various particles by making use of the relationship $E_0 = mc^2$ for each particle. We find then that

$$Q = (m_1 + m_2 - m_3 - m_4)c^2 = \Delta mc^2 \tag{5}$$

The energy difference is now seen to be directly related to the difference in rest-mass between the incoming and outgoing particles.

Now rest-masses can be measured with great accuracy by means of the magnetic mass spectrometer. Mass values good to one part in ten million are now available. Such high precision is possible because the method makes a direct comparison between the unknown mass and the mass of a nucleus such as carbon-12, which is used as a standard. Using Equation (5), then, the Q of a reaction can be measured with equally high precision by using the accurately known mass values of the particles entering into and leaving the reaction, assuming that the mass-energy relationship is valid and also that energy is conserved.

On the other hand, if we want to verify conservation of energy together with the mass-energy relationship, we can measure the kinetic energies of the particles going into and coming out of a reaction as accurately as possible, and use Equation (4) to find the Q of the reaction. If this Q value equals the Q value found by using the masses, then both energy and mass are conserved and the relationship between them is verified.

Notice that we cannot verify conservation of energy merely by measuring the *kinetic* energies of the incoming and outgoing particles with the magnetic analyzer. The reason is that the kinetic energies before and after the reaction are not equal; it is the *total* energies that are equal. The total energy includes the rest-mass energy. That is, in order to find out how much energy was given up by the nucleus in going through the reaction it is necessary to measure the mass lost by the nucleus. The mass-energy relationship is seen to be an integral part of the conservation of energy measurement.

The earliest comparisons of nuclear Q values and nuclear mass differences were made by K. T. Bainbridge, one of the pioneers in precision mass-spectrometry, in the early 1930s. A recent measurement[9] of the mass difference between the isotopes Fe^{57} and Fe^{56} was compared with the mass change during the reaction

$$H^2 + Fe^{56} \longrightarrow Fe^{57} + H^1$$

where H^2 represents a deuteron (a nucleus of deuterium) and H^1 represents a proton. The Q value for this reaction is 5.422 ± 0.006 MeV, which is equivalent to 8.686×10^{-13}

Joule. The mass difference Δm (as defined by Equation 5) represents an energy of 8.688 ($\pm$ 0.0027) $\times$ 10^{-13} Joule. The ratio of Q to Δmc^2 is thus seen to be

$$\frac{Q}{\Delta mc^2} = 0.9998 \pm 0.0012$$

The energy release in this reaction is seen to equal the mass difference to an accuracy of two parts in ten thousand. However, the accuracy of the measurement is only twelve parts in ten thousand.

The largest source of error in this kind of comparison is the Q value measurement itself. Such measurements are not as accurate as mass spectrometer measurements, because in the mass spectrometer you can make a direct comparison of the masses of two kinds of nuclei being deflected through the same magnet simultaneously. In a nuclear scattering experiment the incident particle and the scattered particle are analyzed in two separate magnets, and there arise problems concerning the intercalibration of these two magnets. Therefore typical Q-value measurements have accuracies of one part in a thousand, or 0.1 percent.

However, Taylor and Wheeler[10] have shown that such Q-value measurements can be combined with more accurate mass-difference measurements to make a verification of energy conservation that is good to about two parts in 10^6. This feat is performed by an analysis of an experiment involving a deuteron combining with another deuteron to form a nucleus of tritium (H^3) plus a proton (H^1):

$$H^2 + H^2 \longrightarrow H^3 + H^1 + Q$$

In the actual experiment[11] the kinetic energy of the incident deuteron was measured to be 1.808 $\pm$ 0.002 MeV, the target deuteron was at rest, and the outgoing proton's kinetic energy was found to be 3.467 $\pm$ 0.0035 MeV, at an angle of 90°. The energy of the tritium nucleus was not measured in this experiment, so that the Q-value cannot be obtained directly from these data. However, if it is assumed that conservation of momentum is obeyed, then the Q-value can be calculated (as shown in Appendix II). If this energy agrees with the mass difference, then conservation of momentum and of energy are both verified together.

A different approach is taken by Taylor and Wheeler, who take the measured kinetic energy of the incident deuteron and the energy of the proton coming out of the D-D reaction. These energies are combined with the highly accurate mass values for these particles obtained from mass spectrometer measurements. As shown in Appendix II, this information allows us to calculate the mass of the tritium nucleus, and it is found to be

$$m_3 = 3.016056 \pm 0.000015 \text{ atomic mass units}$$

This may now be compared with the tritium mass obtained directly from the mass spectrometer measurement:

$$m_3 = 3.0160494 \pm 0.0000007 \text{ atomic mass units}$$

These two masses—one obtained from a nuclear reaction, and one measured directly—differ by only about two parts in a million.

This procedure gives a result more accurate than the one part in a thousand error quoted for the kinetic energy measurements themselves, and it seems remarkable that we can draw a conclusion more accurate than the data going into it. The reason for this is seen when we examine the arithmetic used in coming to the conclusion, and we find that the procedure is cleverly chosen so that the kinetic energy values are only small corrections to the atomic masses, which are known to a very great degree of accuracy. In this way, the accuracy of the mass measurements allows us to verify conservation of energy, conservation of momentum, and the Einstein mass-energy relationship all in one fell swoop. (Of course, according to the relativistic point of view, energy, momentum, and mass are all intimately related quantities, so that we are not really dealing with three separate laws, but only one fundamental relationship.)

4. Creation and Annihilation of Particles

One of the fundamental discoveries of elementary particle physics is the fact that for each kind of particle there is a corresponding antiparticle. The familiar "normal" particles are those that make up ordinary matter in our universe. (At least in our solar system; there is a conjecture that galaxies in other parts of the universe may consist of antimatter—matter consisting of antiparticles.) These are the electrons, protons, neutrons, various kinds of mesons, baryons, and so on. Antiparticles may be produced under proper conditions, to live for a brief moment before disappearing in a flash of radiation.

Antiparticle production often occurs in certain types of radioactivity. An atomic nucleus normally has a few more neutrons than protons in order for the nuclear binding force to overcome the electrostatic repulsion of the protons. If a radioactive nucleus (formed by fission or other nuclear reaction) has too many protons for stability, it is possible that one of the protons can turn into a neutron, giving off a charged particle that is just like an electron as far as mass is concerned, except that it has a positive electric charge rather than a negative charge. This positron was the first of the antiparticles to be discovered experimentally, and its existence was a great success for the theory of Paul Dirac, who had predicted that such a particle ought to exist as a result of combining the principles of relativity with the laws of quantum mechanics.

Subsequently it was found that antiprotons could be created by bombarding a metal target with a beam of protons accelerated to energies greater than six billion eV by means of a proton synchrotron. As each high-energy proton interacted with the target nuclei, its kinetic energy was converted into the rest-mass of a proton and an antiproton. This discovery verified the fact that the positron was not just an isolated case. Since that time the antiparticle corresponding to every kind of particle has been created and identified. In

The Conservation Laws: Modern Experiments

general an antiparticle has the same mass as the corresponding particle, but the opposite electric charge. If the particle is neutral (as in the case of the neutron) the antiparticle is also neutral, but the magnetic field arising from its spin motion is reversed, as though opposite charges were whirling about inside the particle.

The creation and annihilation of particle-antiparticle pairs are unique examples of complete conversion of energy to matter and vice-versa. For example, if high-energy gamma rays (from a radioactive source, or from an accelerator such as a betatron) interact with a piece of matter, one of the reactions that can take place is the process of pair-production: a gamma ray photon comes into the vicinity of an atomic nucleus, the photon disappears, and in its place are found an electron and a positron. In order for this reaction to take place, the gamma ray photon must have at least 1.022 MeV of energy, for it must supply the rest-mass energy of both the electron and the positron, each of which amounts to 0.511 MeV.

The converse reaction can also take place. The positron created by pair production (or any other means) wanders along through space or through solid matter, encountering numerous atomic particles along the way, and gradually slowing down. Eventually it makes a head-on collision with an electron, and the two of them disappear in a burst of electromagnetic energy. As a result, two gamma ray photons appear, traveling in opposite directions. The fact that two photons must appear, rather than a single one, is another piece of evidence that momentum is conserved in this type of interaction. Suppose that an electron and positron, traveling very slowly, annihilate each other. The rest-mass of each of the particles corresponds to an energy of 0.511 MeV, so that if a single photon resulted we would expect it to have 1.022 MeV energy and a corresponding amount of momentum. But since the electron and proton, moving slowly, had essentially zero momentum before the reaction, the production of a single photon with a finite momentum would violate the law of conservation of momentum.

In order to conserve both energy and momentum, the result of the annihilation reaction must be two photons, each with energy 0.511 MeV, moving in exactly opposite directions. At first glance we might guess that this should be a very accurate way of verifying the conservation laws. We know the electron and positron rest-masses very accurately; all we need to do is to measure the photon energies and directions with equal precision.

This experiment is done by putting two gamma ray detectors on opposite sides of a radioactive source in which positrons are being produced and annihilated.[12] The detectors send their signals into a coincidence circuit that counts only the photons emerging simultaneously from the source in opposite directions. However, when one of the counters is moved a small distance one way or the other (Figure 6-7), it is found

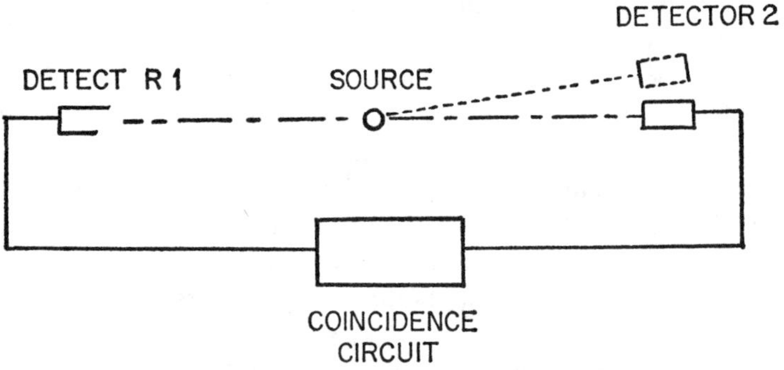

Figure 6-7. The radioactive source produces positrons that are annihilated within the source. The two gamma ray photons produced by the annihilation go off in opposite directions and are detected by the scintillation detectors. The coincidence circuit makes sure that only those gamma rays are detected that left the source simultaneously. If one of the detectors is moved through a small angle, the number of counts is reduced.

that there is a small but definite spread in the directions of those photons. Some of the photon pairs go off in directions

The Conservation Laws: Modern Experiments

that are not exactly opposite, and the average deviation from 180 degrees is about half a degree.

It was quickly realized that the reason for this discrepancy is the fact that the atoms within the solid matter of the source have a thermal motion that corresponds to about 0.025 eV of energy at room temperature. The electrons and positrons, as they come together to annihilate each other, have a minimum energy of this order of magnitude. Therefore the momentum of the interacting particles need not be exactly zero, and as a result of conservation of momentum the two gamma rays need not go off in exactly opposite directions. As far as energies are concerned, the best measurements of the annihilation photon energy is 0.51079 ± 0.0006 MeV, which differs from the electron rest-energy (0.51098 MeV) by 0.04 percent. Gamma ray energy measurements of this accuracy are difficult to come by, so that this kind of verification is the best that can be expected.

In practice, the annihilation phenomenon is not used as a test of the conservation laws; rather, conservation of energy and momentum is assumed, and the observation of the angular spread of the emitted photons is used to study the motions of the atoms within the material in which the positrons are annihilated.

The emission of electrons (beta rays) and positrons from atomic nuclei is a process that, on the surface, appears to violate conservation of energy. This is evident if you measure the energy of the beta rays being emitted by a radioactive isotope, and plot a curve showing the number of electrons appearing with a given energy (Figure 6-8). This measurement may be done by means of a magnetic spectrograph that bends the paths of the electrons in a magnetic field. An isotope extensively studied early in the game was radium E. It was found that the most energetic electrons had an energy of 1.17 MeV, but that most of the electrons reaching the detector had lesser amounts of energy.

The observation that struck everybody most forcefully was the fact that electrons appeared to be emitted with *different*

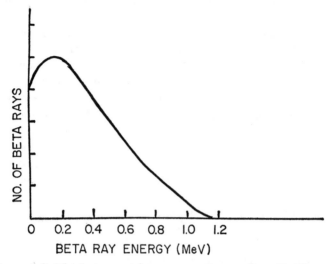

Figure 6-8. The beta-particle spectrum from radium E. The curve shows the number of beta rays (per unit interval of energy) plotted as a function of the energy.

amounts of energy. Yet, if a nucleus with a definite amount of energy shoots out an electron, ending up as a nucleus with another definite amount of energy, all the electrons should have the *same* amount of energy. Thus, if radium E always disintegrates to polonium, and the amount of energy available for this transformation is 1.17 MeV, then all the electrons should have that amount of energy.

At first it was guessed that perhaps all the electrons started out with the same amount of energy, but that in making their way out of the atoms the electrons lost some of their energy. If that theory were true, the missing energy should show up in the form of heat. However, measurements of the heat given off by radium E in its decay showed that only about 28% of the 1.17 MeV energy could be accounted for. It appeared that most of the energy released by the radium E nucleus as it decayed to polonium was transformed into some invisible form that was irretrievably lost. The same

The Conservation Laws: Modern Experiments

mysterious loss of energy was found in every case of radioactivity involving the emission of electrons or positrons.

This observation left the physicists of the 1930s in the uncomfortable position of admitting that conservation of energy was violated, or postulating that energy was being removed from the entire observable system in such a subtle way that it could slip through all the instruments without a trace. Rather than admit the possibility of nonconservation of energy, as well as of momentum and angular momentum, Wolfgang Pauli postulated in 1931 that every time an electron or positron was emitted from a nucleus during a beta decay reaction another particle was also emitted. This new particle had to have a very small—perhaps zero—mass, and it had to have no electric charge, although it was required to have a spin angular momentum with a resultant magnetic field. Because of its lack of charge and negligible mass it could slip through the detecting instruments without leaving a noticeable disturbance. Enrico Fermi named this particle the *neutrino*, or "little neutral one" (it does not mean an Italian neutron), and developed a mathematical theory that was able to predict the shape of the beta ray spectrum—the curve describing the number of electrons of a given energy given off during beta decay.

In addition to the energy measurements, a number of experiments were done in which the angle between the emitted electron and the recoiling nucleus was measured.[13] If the electron was the only particle emitted from the nucleus, then the two should fly apart in opposite directions. However, if a neutrino is emitted at the same time, it must carry with it an amount of momentum proportional to its energy, even though it has no rest-mass (the momentum $p = E/c$, just as in the case of a photon of light). As a result, the recoil of the nucleus will be in some direction that is not opposite to the direction of the electron's travel. The results of the experiments all agreed with the hypothesis that energy and momentum were being conserved and that both energy and momentum were being carried off by an undetected particle.

Finally, in 1956 a group of scientists at Los Alamos succeeded in detecting the elusive neutrinos directly.[14] The very existence of the neutrino, once established, was strong evidence that energy was indeed being conserved. Although there is no way to measure the energy of a neutrino with any degree of accuracy, the overall agreement between the theoretical spectrum shape and the experimental curves leads everyone to believe that the theory is internally consistent.

The importance of demonstrating conservation of energy for beta decay lies in the fact that this particular kind of reaction is caused by a type of nuclear force completely different from the interactions we have been discussing up to now. While we have shown that conservation of energy is good to one part out of 10^{15} as far as the electromagnetic interaction is concerned, we must remember that beta decay is accounted for by the "weak nuclear interaction"—an entirely new kind of force that determines what happens during interaction between electrons and neutrinos. This puts us into a whole new ball game, requiring fresh confirmation of the conservation law.

The success of conservation of energy in all areas of science, in relation to all the forces we can perceive, leads us to the hypothesis that the concept of energy represents a physical phenomenon that cuts across all categories and is common to all kinds of forces. Indeed, recognizing that conservation of energy represents a fundamental symmetry of space-time (the equations of motion are not affected by the particular time we choose to set all the clocks to zero), it is not surprising that if energy is conserved for one kind of force it is then conserved for all the different kinds of force. However, in this game of physics we must always keep our eyes open for surprises.

CHAPTER 7

The Principle of Relativity

1. *Types of Relativity Experiments*

Einstein's Special Theory of Relativity is to modern physics what Newton's laws of motion were to classical physics. Relativity is the physicist's way of viewing the geometry of space and time as well as the relationship between matter and energy. While the word "theory" sometimes has the implication of a conjecture or hypothesis, by now the Theory of Relativity has been so thoroughly tested that it has become a full-fledged conceptual structure on which all of physics is based and against which other theories are tested.

So many experiments have been performed to verify the Principle of Relativity that it would be impossible for me to discuss them all in any detail. Furthermore, there have been so many excellent books describing the basic ideas and consequences of relativity that another addition to this literature would be superfluous. In this chapter I would like simply to examine the fundamental question of what kinds of experiments are necessary to verify the Principle of Relativity, and to describe a few of the most recent and accurate measurements.

For those interested in further explorations into this vast subject, there are a number of extensive bibliographies available.[1] Resnick[2] provides an unusually complete listing of the major relativity experiments, while on a more advanced level D. I. Blokhintsev[3] has surveyed the verifications of relativity obtained by high-energy experiments.

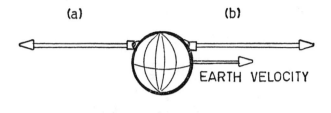
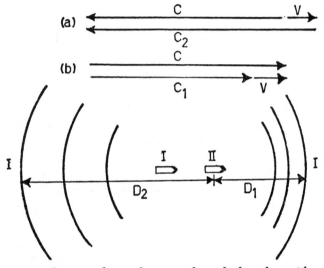

Figure 7-1. Suppose the earth moves through the ether with velocity v, while a beam of light projected from the earth travels with a constant velocity c relative to the ether. In (a) the earth is moving away from the direction of the light beam, so that to the observer on the earth the light appears to go faster than c. In (b) the earth is moving toward the light beam and tends to catch up with it, so that the earth-bound observer sees the light traveling with a velocity less than c. The vector diagrams show how the velocities add up in either case. The situation is similar to that shown in the diagram below, where a boat on a pond moves from left to right producing ripples periodically. The wave sent out when the boat was in position I reaches the position shown by the time the boat reaches position II. Relative to the boat, the forward-moving ripple has gone a distance D_1, while the backward-moving ripple has gone a greater distance, D_2. Therefore the man in the boat says that the backward-moving wave is traveling faster than the forward-moving wave. The fact that the ripples become bunched together in the forward direction and are spread farther apart in the backward direction illustrates the Doppler effect.

The Principle of Relativity

The fundamental question of relativity is whether or not there exists a frame of reference that may be considered absolutely at rest. Newton assumed that there was such a unique frame. Einstein, on the other hand, claimed that there is no special reference frame absolutely at rest, and that as a result no inertial reference frame can be distinguished from any other inertial reference frame by experiments done entirely within that frame. There is no unique frame with special properties.

Historically, the controversy came to a head in connection with the theory of the propagation of light. Nineteenth-century physicists visualized light to be some sort of mechanical wave carried by the *ether*, that famous and elusive fluid through which the planets made their way, and whose sole purpose was to propagate light from one place to another. If the ether theory is correct, then the ether itself represents that unique Newtonian absolute reference frame, and the speed of light should be a constant with respect to that frame, just as the speed of sound is a constant relative to the atmosphere. In no other frame would light travel with constant speed regardless of the direction of propagation.

Suppose that theory were true. Then the speed of light as measured by an observer on the earth moving through the ether should depend on the direction of the light beam relative to the direction of the earth's motion. For example, let the speed of light relative to the ether be c, while the velocity of the earth relative to the ether is v. Then when the light beam is sent ahead of the earth, its velocity relative to the earth is $c_1 = c - v$. On the other hand, if the light beam is pointed in the direction opposite to the earth's motion, then $c_2 = c + v$ is its velocity relative to the observer on the earth (Figure 7-1). It would appear from this that a straightforward set of measurements of the speed of light relative to the earth would allow one to measure the velocity of the earth through the ether and settle the controversy once and for all. For, from the above definitions it follows that $v = \frac{1}{2}(c_2 - c_1)$.

Discovering the Natural Laws

As a matter of fact, the speed of light has been measured numerous times, and is now known to a quite fantastic degree of accuracy: 299,792.5 ± 0.3 kilometers per second. Since the earth goes around the sun with a speed of about 30 kilometers per second, you might expect to find variations in the measured speed of light amounting to plus or minus 30 km/sec—if the ether theory were true. This kind of variation is not found, and there is a strong temptation to use this constant value of the speed of light as evidence that the speed of light is absolutely constant regardless of the motion of the source or receiver, and that therefore the ether theory of light is incorrect.

However, if we apply this sort of plausible reasoning, we fall into a trap. We cannot use the measured value of the speed of light to disprove the ether theory. The reason is that almost all methods of measuring the speed of light involve beams of light (or radio waves) moving back and forth in opposite directions, and so the speed actually measured is an *average* speed for waves making a complete round trip. Recall, for example, that in Michelson's measurement of the speed of light, the light beam is reflected from a rotating mirror to a distant mirror, and then back to the original rotating mirror. As we show in Appendix III, even if the ether theory were true, the average back-and-forth measured velocity $\bar{c}$ would be related to the "true" velocity c by the expression: $\bar{c} = c(1 - v^2/c^2)$.

The *change* in the measured speed of light caused by the earth's motion through the hypothetical ether is seen to be proportional to the square of the small quantity v/c. We would have to know the speed of light to within an accuracy of 0.003 km/sec in order to detect an "ether drift" by this means.

Is there no way to measure the speed of light traveling in one direction? As a matter of fact, there is one such way: the first measurement of the speed of light ever made. When Ole Roemer, in 1676, observed the time it took for Jupiter's satellites to make one orbital revolution, he found that this

The Principle of Relativity

period became shorter while the earth was approaching Jupiter, and became longer while the earth receded. We now recognize that what Roemer saw was essentially the Doppler effect: the change in period (or frequency) of a signal because of the motion of the receiver relative to the source.

We could, in fact, make use of this effect in an up-to-date measurement of the speed of light. For example, suppose we measure the frequency of an electromagnetic wave such as a radar beam with the receiver moving toward the source, and then compare this frequency with the one obtained when the receiver moves away from the source. Usually we use the Doppler shift to find the velocity of the receiver relative to the source, assuming that we know the speed of light. But there is nothing to prevent us from turning the equations around so that if we know the velocity of the receiver relative to the source we can then find the speed of light. Actually, if the ether theory is correct, there are two quantities to find: the velocity of the wave relative to the ether, and the velocity of the source relative to the ether. In theory, at least, both of these velocities can be found with the experiment described above.

In practice, we run into the problem of experimental accuracy. To a first order of accuracy, the speed of light can be found by a single Doppler shift measurement, where you simply measure the change in frequency of the signal when the receiver is in motion. This requires measurement of a *first difference*—the difference between the normal and the shifted frequencies. However, in order to find the effect of motion through the ether, it is necessary to measure a very tiny *second difference:* the difference between the two Doppler shifts—one found with the receiver moving toward the source, and the other with the receiver moving away.

This appears to be a general rule: The speed of light itself can be measured by a "first order" experiment, but in order to detect an ether-drift—a motion of the earth through the ether—the experiment must entail a "second order" measurement, one capable of measuring quantities as small as $(v/c)^2$ —that is, one part in one hundred million.

A great many such experiments have been performed, and the net result is that no motion of the earth through the ether has been detected. The first and most famous of these experiments was the Michelson-Morley experiment, performed at the Case Institute of Technology in Cleveland, Ohio, during the years just prior to 1887.[4] The aim of this experiment was to detect a difference in the speed of light beams traveling in two different directions, as predicted by the ether theory. Since this experiment gave a negative—or "null"—result (i.e., there was no difference in speed), the ether theory of light was thrown into serious doubt.

Einstein later leaped into the breach with an alternative hypothesis—his Principle of Relativity—in which light, by virtue of its nature as an electromagnetic wave, automatically traveled with constant speed relative to any observer in any inertial reference frame. Now Einstein may or may not have been aware of the specific results of the Michelson-Morley experiment. At least he does not refer to it in his famous paper of 1905.[5] What he does comment on is the fact that Maxwell's equations—which deal with the behavior of electromagnetic fields—predict a constant speed of light. Furthermore, he notes that the current induced in a wire moving near a magnet depends only on the relative motion of the wire and magnet, and not on the motion of either one with respect to some outside reference frame. He then goes on to say: "Examples of this sort, together with the unsuccessful attempts to discover any motion of the earth relatively to the 'light medium,' suggest that the phenomena of electrodynamics as well as of mechanics possess no properties corresponding to the idea of absolute rest."

From this springboard, Einstein leaped, with magnificent imagination, to the two great hypotheses—which he labeled postulates—that are the foundation of the Special Theory of Relativity. These well-known postulates are, stated simply:

The Principle of Relativity

1. The laws of physics are the same in all inertial reference frames. No preferred inertial system exists.

2. The speed of light in free space has the same value in all inertial systems. That is, any observer moving with constant velocity will measure the speed of light to be the same, regardless of his motion or the motion of the source.

From these two fundamental postulates Einstein proceeded to deduce many of the results that have since become familiar:

1. A moving rod appears to be shortened in the direction of its motion, so that if its length when at rest is L_0, its length in motion is given by the equation $L = L_0 \sqrt{1 - (v/c)^2}$. This effect is called the Lorentz-FitzGerald contraction, because it was postulated shortly before Einstein's theory by Lorentz and FitzGerald as an attempt to explain away the results of the Michelson-Morley experiment.

2. An object which has a mass m when at rest, is seen to have an increased mass M when in motion, given by the formula $M = m/\sqrt{1 - (v/c)^2}$. As we have seen in the last chapter, this relativistic mass increase is now taken for granted as an everyday event in nuclear and elementary particle physics, and is associated with the fundamental equivalence of mass and energy: $E = Mc^2$.

3. A moving clock appears to run more slowly than the same clock at rest. That is, any kind of oscillator operating with a frequency f_0 when it is at rest relative to the observer is seen to be oscillating with a reduced frequency $f = f_0\sqrt{1 - (v/c)^2}$ when it is moving with velocity v relative to the observer. This effect is called the relativistic time dilation.

The above are the major dynamic effects of relativistic motion. In addition there are the transformations in electric and magnetic fields caused by the motion of the observer. What appears to be an electric field to one observer changes partly into a magnetic field to a moving observer, and vice versa. Thus we conclude that there is no sense in talking about electric and magnetic fields as separate objects, but that they are both simply separate components of a single entity: an electromagnetic field.

In the years following Einstein's original paper, numerous experiments were performed to verify the Theory of Relativity, and a number of controversies arose concerning the interpretation of these experiments. It became clear that the experimenter had to work hand in glove with the theorist in order to be certain that the experiments proved what they were supposed to prove. For example, the Michelson-Morley experiment is sometimes cited as verifying Einstein's relativity principle by showing that the speed of light is a constant. However, as we will see, a careful analysis shows that the Michelson-Morley experiment by itself is not sufficient to prove the constancy of the speed of light.

As was demonstrated by H. P. Robertson, of the California Institute of Technology,[6] Einstein's Principle of Relativity requires at least three different kinds of experiments to test its implications completely, and to eliminate the possibility that some kind of ether theory gives just as satisfactory an explanation as Einstein's theory. The three kinds of experiments described by Robertson are designed to test the following hypotheses:

1. The total time required for light to traverse, in free space, a given distance and then to return to its origin is independent of the direction of travel of the light beam.

2. The total time required for a light beam to traverse a closed path in an inertial frame of reference is independent of the velocity of this reference frame. Or more simply, the velocity of light around a closed circuit is independent of the motion of the source and receiver.

3. The frequency of a moving atomic light source or radio transmitter is altered by the factor $\sqrt{1 - (v/c)^2}$, where v is the velocity of the source with respect to the observer.

All together these three statements are equivalent to the single statement that the speed of light is a constant and is the same in every inertial reference frame. However, breaking it down into three pieces gives us three real experiments

The Principle of Relativity

that can be performed, considering the fact that our measurements on the speed of light generally involve a round-trip motion of the light beam.

Notice that these are not the only kinds of experiments that can be done. Beginning with Einstein's two postulates one can deduce numerous observable consequences—those we have listed above and many more. All of these consequences have been verified by experiment. However, the three crucial experiments listed above have great historical importance. We may consider them to be the experiments that provide the minimum information necessary to verify Einstein's Principle of Relativity.

Most mathematical discussions of the Theory of Relativity begin with the Lorentz transformation equations. These are the equations that tell us where an event is located in one reference frame if we know where it is located in another frame moving relative to the first frame. For example, if you hold a clock in your hand while you are in a moving spaceship, an observer on the ground can describe the position of that clock relative to the earth by using a transformation equation. Furthermore, even though his clock and your clock may have been synchronized before you left the ground, he finds that the two clocks are no longer together as the ship travels away from the earth. Thus, there is also a transformation equation that tells how the time dimension depends on the position and velocity of the spaceship.

These equations are called the Lorentz transformations rather than the Einstein transformations, because Hendrik A. Lorentz actually arrived at them a year or two before Einstein did. However, Lorentz derived them in connection with the special problem of how electromagnetic fields appear to observers in different frames of reference. Einstein, on the other hand, made the problem completely general, applying to all kinds of measurements. Beginning with the postulate that the speed of light is a constant, he showed that the Lorentz transformations were the proper equations to use in going from one coordinate system to another.

Discovering the Natural Laws

Another way to approach the problem, however, is to start with the three fundamental experiments described above. If nature behaves according to those three statements, it can be shown that the Lorentz transformations are the proper equations to use in describing events in space-time. Furthermore, if the Lorentz transformations are true, it follows that light travels with the same speed in all inertial reference frames. The logic is a bit more complicated than simply postulating at the beginning that the speed of light is a constant and deriving the Lorentz transformations from that. It does, however, clarify the individual roles of each experiment.

Once the Lorentz transformation equations are established, then by using the laws of conservation of energy and momentum we can deduce all the other physical consequences of relativity: the increase in mass of a moving object, the equivalence of mass and energy, the behavior of electromagnetic fields, and so on. For these reasons it is worthwhile to consider the three prototype experiments in some detail.

2. The Three Classical Experiments of Relativity

The first of the three statements concerning the speed of light is tested by the Michelson-Morley experiment and by all the similar erperiments that followed it. (We will call such experiments M-M-type experiments.) The essence of such an experiment is shown in Figure 7-2. A beam of light is split into two parts by a half-silvered mirror. One of the two resulting beams goes straight ahead until it is reflected back on its path by a second mirror. The other beam is turned at right angles and travels until turned back by a third mirror. The two beams of light that started out in phase with each other travel over different paths in different directions, and then return to be focused together onto a photographic plate. The result is a set of interference fringes. Those parts of the beam that traveled exactly the same distance—or whose paths differed by an integral number of wavelengths—join together in phase and add onto each other, forming a bright fringe. Those parts of the beam whose path lengths differed by half

The Principle of Relativity

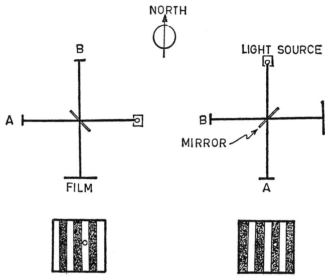

Figure 7-2. A schematic diagram of the Michelson-Morley experiment. A beam of light from a monochromatic source is split into two parts by a half-silvered mirror. Part of the light goes through the mirror and is reflected back on its path by a second mirror at A. The part of the light reflected from the half-silvered mirror goes to mirror B and is reflected so that the two beams join together once more at the half-silvered mirror and go on to the photographic film, where they leave their record. Those parts of the light beams whose path lengths differed by half a wavelength will cancel each other out, leaving a dark fringe. When the apparatus is turned through ninety degrees, the number of wavelengths in each of the paths will change—if the light speed depends on the direction of propagation. This will appear as a movement to one side or the other of the interference fringes. In the drawing we have shown a shift of half a fringe. (One dark plus one light fringe equals a whole fringe.)

a wavelength will undergo destructive interference and will form a dark fringe. With the mirrors adjusted properly, a series of alternating light and dark fringes can be photographed.

With this arrangement alone, you could not tell whether

Discovering the Natural Laws

one beam was traveling with a different speed than the other, because you do not know the lengths of the two arms sufficiently accurately. To find a difference in speed you must now rotate the apparatus through an angle of ninety degrees. Then the beam that goes in the east–west direction now hits mirror B, while previously it went to mirror A. At the same time the north–south beam now goes to A.

Suppose there is something about the directions in space that makes an east–west beam travel faster than a north–south beam. With the apparatus in position I, the east–west beam will proceed via mirror A, and a given crest on this wave will arrive at a particular point on the photographic plate a short time before the beam coming from mirror B. (For simplicity, let's suppose that this difference in time amounts to a quarter of a cycle.) Now the apparatus is turned to position II, and the light from mirror A, going north–south, reaches that same point on the photographic plate a quarter of a cycle later than the light from mirror B. There is a total change of half a cycle in the phase of one beam with respect to the other. The result of this is that where there was constructive interference at some particular point on the plate while in position I, in position II there is now a dark fringe. In other words, the entire interference fringe pattern has moved over by half a fringe. Of course, if the difference in light speed between the two beams is less than assumed in the example above, the amount of fringe shift will be less.

The original Michelson-Morley experiment was sensitive enough to detect a shift of 1/100 fringe, and within that limit of experimental accuracy no fringe shift was found. Recall that the reason for expecting a fringe shift was because of the prevailing theory of light transmission through the ether. According to that theory, if the earth is traveling through the ether, then one of the beams will be going up- and downstream, while the other beam will be traveling cross-stream. The cross-stream beam should appear to go more rapidly than the other beam, and so there should be a noticeable fringe shift upon the successful operation of the

The Principle of Relativity

M-M experiment. (See Appendix III for details of the mathematics.) If we put the 30 km/sec orbital velocity of the earth into the equation, we find that there should be a shift of four-tenths of a fringe. Thus the accuracy of the original M-M experiment was sufficient to measure an ether velocity one-fortieth of that minimum amount expected. (Actually, one might expect an even greater effect because of the motion of the sun and the entire solar system through the galaxy.)

The simplest explanation of this null result—simple to our modern eyes—is that the speed of light is nothing more than a constant, and so is not affected by the motion of the earth. However, this straightforward explanation was not satisfying to those who thought that light ought to behave like any other mechanical wave propagating through a medium. Therefore an alternate hypothesis was proposed that allowed the ether theory to be retained, but accounted for the null result of the M-M experiment.

This alternate hypothesis (proposed before Einstein by George F. FitzGerald and Hendrik A. Lorentz) started with the ether theory of light, which predicted, as previously described, that the light going in the direction of the earth's motion through the ether would be seen to travel more slowly than the light going at right angles. Lorentz and FitzGerald postulated that the interferometer arm pointing in the direction of the earth's motion would be shortened as a result of its motion, while the other arm would be unchanged. The amount of shortening was assumed to be just enough to compensate for the change in light speed, so that as a result the light took the same amount of time to traverse both arms, and no phase shift was observed.

It turns out that to explain the zero fringe shift by the contraction of the interferometer arm, the length of the contracted arm must be given by the equation $L = L_0\sqrt{1 - (v/c)^2}$, where L_0 is the length of the arm when it is at rest (or pointed at right angles to the direction of motion), and L is its contracted length when it is pointed in the direction of the motion.

Discovering the Natural Laws

Interestingly enough, this Lorentz-FitzGerald contraction is exactly the same contraction that is predicted by Einstein's Theory of Relativity starting with the hypothesis that the speed of light is a constant. For this reason the M-M experiment, by itself, is not able to distinguish between Einstein's relativity theory and the ether theory with the Lorentz-FitzGerald contraction added to it.

Thus we must look at experiments designed to test the second of the three statements concerning the speed of light. Does the round-trip time for a light signal depend on the velocity of the source as well as the direction of propagation of the light? Perhaps if we repeated a M-M-type experiment at different times during the day, or at different times of the year, we could take advantage of the fact that as the earth rotates around its axis and revolves in its orbit around the sun, the experimental apparatus moves with different velocities relative to the distant stars. Then we might find a shifting of the fringes every twenty-four hours or every twelve months.

Unfortunately, it turns out that the ordinary M-M-type experiment is incapable of observing such an effect, because in such experiments the lengths of the two interferometer arms are made to be equal. The mathematical analysis of the experiment (Appendix III) shows that only by making an interferometer with unequal arms can the effect of the earth's motion on the velocity of light be observed—if such an effect exists.

The first experiment of this type was performed at the California Institute of Technology during the years 1929-31 by Roy J. Kennedy and Edward M. Thorndike.[7] The apparatus of the Kennedy-Thorndike (K-T) experiment was superficially similar to that of the Michelson-Morley experiment, but with certain important differences (Figure 7-3). The arms of the interferometer were not only of different lengths, but they were inclined at an angle less than ninety degrees. Since observations had to be carried out over extended periods of time, it was necessary that the lengths of the arms

The Principle of Relativity

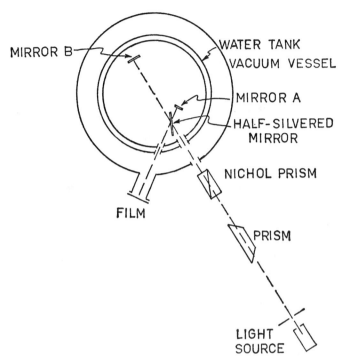

Figure 7-3. The apparatus used in the Kennedy-Thorndike experiment. The light source was a mercury arc; the beam was sent through a prism to select only the single spectrum line of interest, and then was sent through a Nichol prism to polarize the light. The water tank and the vacuum vessel were designed to keep the temperature inside as constant as possible over a long period of time.

be kept from changing due to variations in temperature during that time.

To keep the changes in length down to two parts in 10^{10}, the essential parts of the apparatus were made out of fused quartz, a material that has a very small temperature coefficient of expansion. The interferometer base was a quartz disc 28.5 cm in diameter and 3.8 cm thick, enclosed in a vacuum chamber and temperature-controlled to within 0.001 degree

Discovering the Natural Laws

C. This fine temperature control was accomplished by keeping the chamber inside a thermostatically controlled water tank, which in turn was enclosed within two rooms-within-rooms, each one temperature-controlled.

To perform the experiment, the mirrors were adjusted so that interference fringes were obtained with the 5461 Angstrom line of the mercury spectrum. These fringes were photographed on a plate, the plate was moved a short distance, another set of fringes were photographed, the plate was moved again, and so on, until six exposures had been made, a half hour apart. Twelve hours later the same procedure was followed, but this second set of six pictures was photographed using the spaces between the six pictures in the first set. The result was a set of fringe photographs lined up in a row, with alternating pictures taken twelve hours apart. A microscopic examination of this plate was made to search for fringe shifts due to the rotation of the earth.

In 1931 a search was made for an effect due to the orbital motion of the earth, with sets of photographs taken every three months for nine months. While some minimal amount of fringe shifting was found, the effect could only account for an earth velocity of about ten kilometers per second, which is much smaller than the velocity of the entire solar system through space. Therefore the result of this experiment was considered to be null, with about the same degree of precision as the Michelson-Morley experiment. The striking feature of such an experiment is the vast care and patience and attention to fine detail required to eliminate the possibility of false fringe shifting due to mundane causes such as temperature changes.

Notice that the title of the original paper describing the Kennedy-Thorndike experiment was: "Experimental Establishment of the Relativity of Time." The reason for this title is that a stationary observer looking at the Kennedy-Thorndike apparatus whirling around the sun would explain the absence of fringe-shifting by saying that one of the interferometer arms became shortened in the direction of motion

The Principle of Relativity

according to the Lorentz formula, while at the same time the atomic clocks emitting the light in the mercury sources had their frequencies reduced by the factor $\sqrt{1 - (v/c)^2}$. Thus the Kennedy-Thorndike experiment is a test of the length contraction and the time dilation effect simultaneously.

These are, of course, the effects predicted by the theory of relativity, starting with the basic assumption that the speed of light is a constant regardless of the motion of source or observer. However, we are not quite ready to say that the Kennedy-Thorndike experiment is a complete test of the theory of relativity, because there is one other effect that might be taking place unbeknownst to us—an effect that would be missed by both the Michelson-Morley and Kennedy-Thorndike experiments.

Suppose that the Kennedy-Thorndike apparatus moves past a stationary observer. He sees one of the two interferometer arms (the one pointing in the direction of motion) becoming shorter, while the other is unchanged. He sees the atomic clock frequencies slow down by the time dilation. But suppose that in addition to these changes there is an alteration of *all* the lengths and all of the time scales by some common factor. This factor might depend on the velocity, and it might produce either an increase or decrease in scale length. Whichever it is, such a change could never be noticed by the observer within the moving reference frame, because if he tries to measure the speed of light all his yardsticks and clocks have changed by the same factor. Thus no velocity measurement (a length divided by a time interval) *within* the moving frame can detect this effect.

However, an outside stationary observer would see that the clocks in the light source had changed their frequency by an amount different from that predicted by the time dilation formula. The theory of relativity, based on the constancy of the speed of light in all reference frames, requires that the frequency of a moving clock be given by the formula $f = f_0\sqrt{1 - (v/c)^2}$ and nothing else. Thus, an independent measurement of a clock frequency must be made in order to

Discovering the Natural Laws

be sure that the time dilation is precisely as predicted by Einstein's relativity.

The most direct experiment of this type would be to take two synchronized clocks, leave one at rest in the laboratory, and carry the other around the world in a fast-moving airplane. The moving clock should record a smaller elapsed time than the stationary clock. Until recently, real clocks have not had the precision to detect the very small time difference expected in this measurement. During the past few years, cesium beam frequency standards have been built that are portable enough to use in such experiments, and their precision is just beginning to get to the point where they could detect a relativistic time change. Experiments involving such clocks being transported around the earth in satellites are planned for the future, and their results will be a very direct verification of the Principle of Relativity.

To obtain a more easily detectible change in frequency it is necessary to use a clock traveling at a very great speed. Such a clock can be the atomic oscillator responsible for the emission of light from an excited atom. The wavelength of light emitted from a quantity of ionized hydrogen gas can be measured to a high degree of precision with a grating spectrograph. If the wavelength of the light from rapidly moving hydrogen atoms is compared with the wavelength of light emitted by stationary atoms, it is possible to measure the effect of the changing frequency of the light source, since the wavelength equals the speed of the light divided by the frequency.

Such an experiment was first performed by Herbert E. Ives and G. R. Stillwell at the Bell Telephone Laboratories in New York in 1938.[8] The apparatus consisted of an ion source capable of producing a beam of hydrogen ions of precisely controlled energy. As these ions passed through the hydrogen gas of the tube, and as electrons rejoined the ions to neutralize them, the familiar light of the hydrogen spectrum was emitted from the rapidly moving particles. A mirror arrangement allowed a powerful spectrograph to photograph

The Principle of Relativity

simultaneously the light emitted in the direction of the ion motion and in the reverse direction (Figure 7-4).

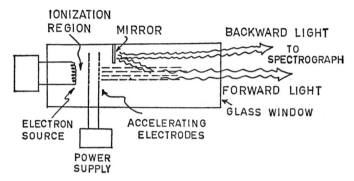

Figure 7-4. The apparatus used in the Ives-Stillwell experiment. The diagram shows a vacuum tube filled with hydrogen gas at low pressure. The filament on the left emits electrons that are accelerated by an electrode so that they ionize some of the hydrogen atoms. The ionized atoms are accelerated by a second electrode so that they form a positive ion beam in the right half of the tube. The light emitted from the hydrogen atoms leaves the tube and is analyzed by a grating spectrograph. Light emitted from the hydrogen atoms in the backward direction is reflected from a small mirror and is sent forward to the same spectrograph.

Now if the classical Doppler shift formula were true in this situation, the wavelength of a given spectrum line would be shifted equally in both directions—toward long and short wavelengths—according to the well-known formula $\lambda = \lambda_0 (1 \pm v/c)$. (Here λ_0 = wavelength of light from the source at rest, λ = wavelength from the moving source, v = source velocity, c is the speed of light, and the $+$ and $-$ signs are for the source receding and approaching, respectively.)

Einstein's relativistic formula for the Doppler shift, on the other hand, predicts that the wavelength from the moving source would be given by a slightly different formula:

$$\lambda = \frac{\lambda_0 (1 \pm v/c)}{\sqrt{1 - (v/c)^2}}$$

Discovering the Natural Laws

If v is much less than c, then the denominator can be expanded by the binomial theorem, and we find that to a second order of approximation the shifted wavelength is given by:

$$\lambda = \lambda_0 \, (1 \pm v/c)(1 + \tfrac{1}{2} v^2/c^2 + \ldots)$$
$$ = \lambda_0 \, (1 \pm v/c + \tfrac{1}{2} v^2/c^2 + \ldots)$$

We see that this formula starts out looking just like the classical Doppler-shift formula, but there is the added term $\tfrac{1}{2} v^2/c^2$, which is always positive, regardless of whether the source is moving toward or away from the observer. Thus, as shown in Figure 7-5, both the direct and reflected rays from

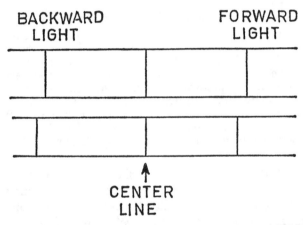

Figure 7-5. One of the lines of the hydrogen spectrum, split into three lines by the Doppler shift. The center line comes from hydrogen atoms in the tube that were essentially at rest, not part of the hydrogen beam. The two lines on each side come from the rapidly moving atoms and demonstrate the Doppler shift: an increased wavelength for the light emitted in the backward direction, and a decreased wavelength for the forward light. If the classical (prerelativistic) Doppler shift were correct, the two shifted lines would be found at equal distances on each side of the center line. The second-order relativistic effect shifts both lines a little bit to the red end of the spectrum, so that one is now closer to the center line than the other.

The Principle of Relativity

the moving hydrogen ions would have their spectrum lines shifted very slightly toward the longer wavelength (red) part of the spectrum.

This effect is, in fact, just what Ives and Stillwell observed in their pictures. However, Ives' interpretation of this result was quite different from Einstein's. Ives connected the shift in wavelength with the changing clock frequency by the following argument: Suppose we consider the theory that the light travels through an ether with a velocity c relative to the ether, while the source is moving through the ether with velocity v. The classical (nonrelativistic) Doppler formula predicts that the observed frequency ν is given by the equation

$$\nu = \frac{c}{\lambda} = \frac{c}{\lambda_0(1 \pm v/c)} = \frac{\nu_0}{(1 \pm v/c)}$$
$$= \nu_0 \, (1 \mp v/c + v^2/c^2 + \ldots)$$

(where ν_0 is the frequency of the clock at rest). Notice that the frequency is greater than the simple quantity $\nu_0 \, (1 \mp v/c)$ by the amount v^2/c^2.

But now make the hypothesis that the atomic clocks in motion oscillate more slowly than the same clocks at rest by the factor $\sqrt{1 - v^2/c^2}$. The frequency of the wave from the moving source is now

$$\nu = \frac{\nu_0 \sqrt{1 - v^2/c^2}}{(1 \pm v/c)}$$
$$= \nu_0 \, (1 \mp v/c + \tfrac{1}{2} v^2/c^2 + \ldots)$$

This quantity is seen to be smaller than the classical shifted frequency by the factor $\tfrac{1}{2}(v/c)^2$, which you would expect from the slowing down of the clocks. If we write this equation in terms of the wavelength, we obtain:

$$\lambda = \frac{c}{\nu} = \frac{\lambda_0 \, (1 \pm v/c)}{\sqrt{1 - (v/c)^2}}$$

which is just the same as the formula obtained from relativity and is just what is found by the experiment. But notice that the theory Ives set out to check with his experiment was an ether theory, not Einstein's relativity.

The most fascinating aspect of this entire matter is the fact that Ives never did believe in Einstein's Theory of Relativity. For a period of several years prior to 1940 Ives wrote a number of papers trying to justify the ether theory of light transmission. He knew that in order to explain the Michelson-Morley experiment it was necessary to assume that measuring rods contracted in the direction of their motion, while in order to explain the Kennedy-Thorndike experiment it was necessary to assume that moving clocks ran more slowly than clocks at rest. Ives' theory was that you could understand everything about light if you assumed that it traveled with constant speed relative to the ether, with the additional provisions that lengths contracted and clocks ran slow. Therefore it was with great courage that he set out to show directly that atomic clocks do indeed run slowly when they are moving.

The ironic ending of the story is the fact that in proving his point Ives also proved Einstein's Theory of Relativity, a theory that Ives emphatically did not want to prove. As we have already mentioned (and as Robertson showed), if we begin with the Michelson-Morley, the Kennedy-Thorndike, and the Ives-Stillwell experiments, we can derive from them the Lorentz transformation equations. These equations include the Lorentz-FitzGerald contraction and the time dilation, which is what the experiments are all about. In addition, we can use these Lorentz transformations to show that if the speed of light is a constant in one frame of reference (the ether, for example), then it follows quite rigorously that the speed of light will turn out to be the same constant in all other reference frames moving with constant speed relative to the first. Thus there is no way of distinguishing between the first frame and all the other frames, and for this reason the ether becomes a superfluous concept.

What we are left with is Einstein's relativity theory, in which the fundamental concept is the idea that light really does travel with the same speed to all observers, and that the only kind of motion we can talk about with meaning is the motion of one object relative to another—absolute motion is out.

In this conceptual structure the idea of the ether is better replaced by the concept of the electromagnetic field. What makes the electromagnetic field more real than the ether is the fact that we can measure electric and magnetic field strengths, while there was never anything we could measure about the ether. The electromagnetic field is therefore a useful concept, while the ether concept never had a use other than psychological.

3. *Modern Experiments*

The development of new optical techniques has permitted the three classical relativity experiments to be repeated with greatly improved precision. The laser, as a source of very coherent and monochromatic light, is a natural candidate for ultraprecise M-M and K-T experiments. In 1964 T. S. Jaseja, A. Javin, and C. H. Townes[9] reported on an experiment in which two laser beams directed at right angles to each other were joined together by an oblique mirror and sent into a photodetector where their interference pattern was picked up in the form of a beat frequency, amplified electronically, and recorded on a chart (Figure 7-6). The entire apparatus was suspended on a platform that was rotated back and forth through an angle of ninety degrees with a period of twenty seconds. For greater sensitivity and better signal-to-noise ratio the detector was tuned to look for any effect occurring with a period of twenty seconds. The sensitivity of the experiment was such that an "ether drift" less than 1/1000 of the velocity of the earth would have been detected, and none was found. To achieve this feat, the instrument had to be capable of detecting a frequency difference of three kiloHertz out of 3×10^{14} Hz, that is, a precision of one part out of 10^{11}.

This laser experiment was an M-M type of experiment, since the wavelength of each laser beam depended on the

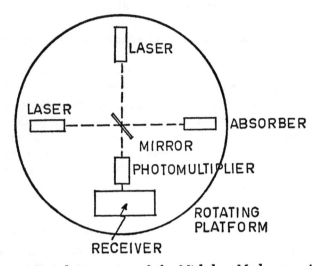

Figure 7-6. A laser version of the Michelson-Morley experiment. Two individual lasers produce light beams that are joined together at the half-silvered mirror and sent into a photomultiplier tube, where any variations in the light intensity are detected. If there is any change in the light wavelength as the platform is rotated back and forth, this will appear as an interference or beat frequency in the output of the photomultiplier tube.

length of the laser tube acting as a resonant cavity. The experiment thus compared the lengths of two laser tubes at right angles to each other.

C. H. Townes, one of the original inventors of the laser, was also involved in a Kennedy-Thorndike type of experiment performed with an ammonia maser. (The maser and the laser are essentially the same type of device, except that the maser operates at microwave frequencies, using molecular vibrations to control the frequency, while the laser works in the infrared and visible part of the spectrum.) In the ammonia maser the frequency of the microwaves produced depends both on the length of the cavity and on the intrinsic frequency of the oscillating ammonia molecules. The experiment performed by J. P. Cedarholm, G. F. Bland, W. W.

Havens, and C. H. Townes[10] was based on the comparison of the frequencies produced by two maser oscillators. The ammonia molecules were concentrated in the form of beams passing along the axis of each cylindrical cavity. The resonant frequency depended on the speed of the molecules relative to the cavity. In addition, if the ether hypothesis were true there would also be a dependence of the resonant frequency on the velocity of the cavity relative to the ether. The two masers, with beams traveling in opposite directions, were adjusted so that the two frequencies were only 20 cycles per second apart—that is, the beat frequency of the two combined signals was 20 Hz. As the apparatus was rotated, one would expect to see a variation in this 20 Hz beat frequency (according to the ether theory), since the velocity of the molecules relative to the ether would be changing. No such effect was noticed, even though a variation of 1/50 Hz out of the 23,870 MHz oscillation frequency could have been detected. The conclusion was that the maximum hypothetical ether drift was less than 1/1000 of the earth's orbital velocity.

An interesting variation of the Ives-Stillwell experiment has been performed using gamma rays instead of visible light, and using the sensitive frequency discrimination of the Mössbauer effect as a detector for the Doppler shift. The experiment of Hay, Schiffer, Cranshaw, and Egelstaff,[11] performed at the Atomic Energy Research Establishment in Harwell, England, made use of the so-called transverse Doppler shift predicted by relativity. In the formula for the Doppler shift discussed previously, the ordinary first-order shift (the term proportional to v/c) depends on the direction of motion of the radiation source relative to the detector. It is greatest when the source is moving directly toward or away from the detector, producing either an increase or decrease of frequency, depending on the direction of motion. If the source is moving at right angles to the direction of motion of the radiation, the first-order Doppler shift disappears. However, there still remains the second-order (or transverse) Doppler shift: a decrease in frequency proportional to $\frac{1}{2}(v/c)^2$, caused by the relativistic time dilation—the slowing down of the time in the moving reference frame. This transverse shift does not depend on the direction of motion at all, but depends only on the speed of the source.

In the experiment of Hay *et al* a radioactive Co^{57} source was mounted at the center of a rotating wheel, while the Fe^{57} absorber was located toward the edge of the wheel (Figure 7-7). The gamma ray detector was stationary, on the outside

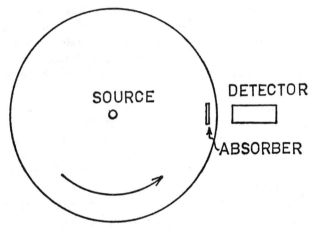

Figure 7-7. A Mössbauer effect experiment to detect the transverse Doppler shift caused by the time dilation. The radioactive Co^{57} source is at the center of a rapidly rotating wheel, while the Fe^{57} absorber is out near the edge of the wheel. A gamma ray detector is stationary, outside the wheel. When the resonant frequency of the absorber is changed by its motion, the number of gamma rays passing through it is increased.

of the wheel, and was gated so as to detect only those gamma rays that entered it when the absorber was between the source and the detector. With the entire wheel stationary, resonance absorption of the gamma rays in the Fe^{57} was observed. But with the wheel rotating, the absorber was now in motion at right angles to the velocity of the gamma rays, and the resonance curve was shifted by virtue of the transverse Doppler effect. As a result, the counting rate obtained by the detector was increased by the amount predicted by calculations based on relativity theory.

Interestingly enough, the original intent of this experiment was to measure the frequency shift caused by the fact that the gamma rays as seen by the rotating absorber are falling through a centrifugal force field, and are expected to experience an energy (and frequency) change—in the same way as photons falling through a gravitational field. The effect is the same as that expected from the transverse Doppler shift theory—a tribute to the consistency of the over-all theory. There is no embarrassment concerning which theory to use. The concept of the centrifugal force field applies if you are sitting on the rotating frame of reference with the absorber, while the transverse Doppler effect is the valid theory if you are stationary in the laboratory frame of reference. Both of these interpretations correctly predict the observed happening.

An experiment outwardly similar to the one just described, but much different in philosophy, was reported in 1963 by D. C. Champeney, G. R. Isaak, and A. M. Khan.[12] In this experiment the Co^{57} source and the Fe^{57} absorber were mounted at opposite ends of a rapidly rotating centrifuge, and there were gamma ray detectors in the north–south and east–west directions. The aim here was not to seek a constant transverse Doppler shift, but instead the experimenters looked for a change in the first-order Doppler shift predicted by the ether-drift theory. If the ether theory is correct, there should be observed a frequency change $\Delta\nu$ equal to

$$\Delta\nu = \nu \mathbf{v} \cdot (\mathbf{u}_a - \mathbf{u}_s)/c^2$$

where ν is the frequency of the gamma radiation, $\mathbf{v}$ is the velocity of the ether past the laboratory, and $\mathbf{u}_a$ and $\mathbf{u}_s$ are the velocities of absorber and source relative to the laboratory. (Note: $\mathbf{u}$ and $\mathbf{v}$ are vectors, and the product in the above equation is defined as $\mathbf{u} \cdot \mathbf{v} = uv \cos\theta$, where θ is the angle between the directions of $\mathbf{u}$ and $\mathbf{v}$.)

In the experiment, as the rotor spins rapidly at a rate of several hundred revolutions per second, the angle between **u** and **v** changes, so that the first thing to look for is a frequency modulation of the gamma radiation. Such a change in frequency will be detected as a change in the resonant absorption, resulting in an increase in the counting rate observed by the detectors. In addition, the magnitude of v should change as the earth rotates on its axis (since presumably v depends on the earth's axial rotation together with its orbital motion). Therefore one might expect to find a daily shift in the gamma ray frequency.

Neither effect was observed, so that this experiment represents another negative result in the effort to verify the ether hypothesis. Actually, of course, most modern experimenters no longer believe in the ether hypothesis. Although I have described these experiments as though they were designed to make a choice between the ether hypothesis and the relativity hypothesis, at this point in history the choice has already been made. The relativity hypothesis is the only one that passes all the tests. However, the ether hypothesis remains a convenient and traditional reference point for comparing the accuracy of the various "null type" experiments.

For example, the most accurate of the classical Michelson-Morley experiments (Joos, 1931) could have detected an ether drift of 1.5 kilometers per second relative to the laboratory. The upper limit on Cedarholm's maser experiment was thirty meters per second, a factor of fifty better. The Mössbauer effect experiment of Cranshaw *et al* was estimated to give an upper limit for ether speed of ten meters per second. The last experiment discussed, that of Champeney and collaborators, is considered to be the most precise of all, claiming an upper limit of 1.6 ± 2.8 meters per second on the possible ether drift.

4. *Other Experiments*

To discuss all the relativity experiments would require an entire volume. I have purposely omitted entire classes of

The Principle of Relativity

experiments, such as those performed with high-energy particles, those relating to the relativistic mass increase, and those having to do with general relativity.[13] In general, I have confined myself to experiments of the three fundamental types needed to establish the validity of the Lorentz transformation equations.

Mass-increase experiments are important because they show that conservation of momentum and energy are valid in all inertial reference frames. Every particle accelerator is designed using the principles of relativistic dynamics; the very fact that they work according to design shows that the Theory of Relativity is an accurate way of looking at nature. The idea that the speed of light is a constant and that the laws of nature are the same in every inertial reference frame are integral parts of the scientist's thinking process. From this point of view the Lorentz-FitzGerald contraction and the time dilation are among the necessary consequences of the nature of time and space, rather than *ad hoc* assumptions patched onto an ether theory of light.

Once we have built up this conceptual structure through the tedious route of experimentation, then we can turn around and *begin* by saying: Of course the speed of light is a constant. It has to be, because the number c is just a constant of proportionality between units of measuring space and units of measuring time in a four-dimensional universe. (This is the starting point of E. F. Taylor and J. A. Wheeler in their imaginative book *Spacetime Physics*.) Now that the skirmishes have been fought and the battles have been won, the scientists have a mental picture of the universe quite different from that prevalent a hundred years ago. Since this new picture is based upon an enormous amount of experimental observation, it stands on a more secure footing than the old model. At the very least, any new theory that comes along must be able to fit in with the same set of observational data.

CHAPTER 8

Electromagnetism

1. *Fundamental Ideas*

The evolution of our ideas concerning electricity and magnetism during the past hundred years is a perfect example of how our theoretical ideas determine the interpretation of an experiment, and how these ideas decide what experiments we consider to be the most important and fundamental. When Heinrich Hertz, in 1888, showed that electromagnetic waves travel with the same speed as light waves, this was considered powerful evidence that light waves were fundamentally the same as electromagnetic waves, differing only in wavelength. Even stronger evidence was the fact that the numerical value of the speed of light was closely related to the electrostatic and electromagnetic units of charge in common use.

The electrostatic unit of charge is defined by making use of Coulomb's Law—the inverse-square law of force between two stationary electric charges:

$$F = \frac{q_1 q_2}{r^2}$$

Here q_1 and q_2 represent the quantity of each electric charge, and r is the distance between them. If we arrange things so that the force between two identical charges is one dyne when the distance of separation is one centimeter, then each of the charges is defined to be one electrostatic unit (esu). (Coulomb's Law will be more fully discussed in part 4 of this chapter.)

The electromagnetic unit of charge is defined by making

Electromagnetism

use of a similar inverse-square law of force that exists between electric currents. If I_1 and I_2 are electric currents passing through small segments of wire, of length L_1 and L_2, then the force between them is given by the equation

$$F = \frac{I_1 I_2 L_1 L_2}{r^2} \sin \theta$$

where the angle θ is as shown in Figure 8-1.

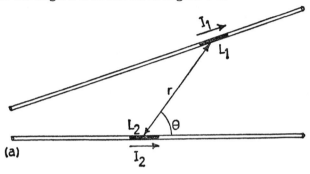

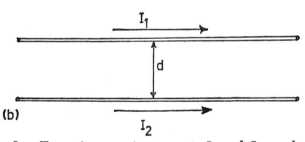

Figure 8-1. Two wires carrying currents I_1 and I_2 are located in the same plane. Consider two small sections of length L_1 and L_2. The force exerted on L_1 by the current in L_2 is inversely proportional to the square of the distance between the current segments, and also depends on the sine of the angle θ, as shown. If the two wires are not in the same plane, then a second angle enters the equation. To find the total force between the two wires, the contributions from all the little wire segments must be summed up. If the two wires are parallel, as in (b), the total force between them varies inversely as the distance between them.

Discovering the Natural Laws

This expression becomes more familiar if we add up the forces due to all the tiny elements in a pair of *parallel* wires, so that we end up with the total force acting on each centimeter of wire:

$$F = \frac{2I_1 I_2}{d}$$

If we make the currents through the two wires equal, if we make the distance d between them one centimeter, and then adjust the current until the force acting on each wire is two dynes, then this measurement defines the current running through each wire to be one electromagnetic unit (emu) of current. (One emu of current is one emu of charge flowing per second.) The formula we have just written down is the Biot-Savart Law relating the magnetic force between two wires to the distance between them.

The question now arises: How big is an emu of charge compared with an esu? Experiments of many kinds showed that the emu was a much larger quantity of charge than an esu—in fact, about 3×10^{10} times larger. That is, it took 3×10^{10} times as many electrons to make up an emu as it did to form an esu, indicating that the electrostatic force from each electron was much stronger than the electromagnetic force. The striking thing about this number is that it appears to be the same as the speed of light (in centimeters per second). Just as striking is the fact that this same number c—the ratio of the emu to the esu unit of charge—also appears in the equation that describes the propagation of electromagnetic waves.

This wave equation in turn is derived from the famous equations developed by James Clerk Maxwell in 1864—the equations that give relationships between changing electric and magnetic fields in free space. According to Maxwell's equations, if you produce a changing electric field (in an antenna, for example), this induces a changing *magnetic* field, which in turn causes a changing *electric* field, and this disturbance in the electromagnetic field propagates away from

Electromagnetism

the antenna with a speed that is equal to c—the ratio of electromagnetic units to electrostatic units.

By measuring the speed of the electromagnetic waves, Hertz was able to verify the prediction of Maxwell's equations, for his waves did travel with the velocity c, and as closely as could be measured this velocity was the same as the speed of light. As a result of these measurements, the number c was seen to be a very significant number—a number that could be measured in two entirely separate ways. It first appeared as the ratio of the emu to the esu unit of charge, and then it turned out to be the speed of an electromagnetic wave. The fact that the two independent ways of measuring c gave the same numerical result was taken as a good validation of Maxwell's Theory of Electromagnetic Radiation, and incidentally gave proof that light was just another kind of electromagnetic wave.

For this reason it was considered important to make measurements of the emu/esu ratio as accurately as possible so that it could be compared with the speed of light. Over the years many such measurements were made, and as of 1933 the most accurate of these measurements (made by Rosa and Dorsey at the National Bureau of Standards) gave a value of $c = (2.9979 \pm 0.0001) \times 10^{10}$ cm/sec, agreeing with Michelson's measurements of the speed of light within the limits of error shown.[1]

In this experiment the emu/esu ratio was determined by measuring the capacitance of a cylindrical condenser. It is known that by measuring this capacitance in terms of the current required to charge it or discharge it, the capacitance in emu is obtained. On the other hand, if the capacitance is calculated from the dimensions of the condenser, one finds the esu capacitance. The ratio of these two capacitances is then found to equal c^2.

Here, then, was a most remarkable result: A measurement involving nothing but the magnitude of an electric current and the geometry of a metal cylinder gave a number equal

Discovering the Natural Laws

to the speed of light. As was clear to any scientist, such a result could not be a coincidence. It must be related to some very fundamental properties of electromagnetic fields.

The most ironic feature of this history is that Maxwell's electromagnetic equations were initially designed with the ether theory in mind, and yet it was not until the ether theory was discarded in favor of relativity that a really fundamental understanding of the connection between the speed of light and the emu/esu ratio could be attained. Just as the Cheshire cat disappeared until nothing was left but the smile, the idea of the ether gradually faded away until nothing was left but electric and magnetic fields. But one important feature was added: the Lorentz transformation equations, which showed how these fields changed their nature when viewed from various frames of reference.

From the modern point of view, a magnetic field is nothing but part of the electric field that becomes apparent when both the charge producing the field and the test charge detecting the field are in motion relative to the observer. From this point of view the factor c enters the equation in just the same way that it makes its way into $E = mc^2$, where it acts as a constant of proportionality between units of mass and units of energy.

To see how this comes about, let us think of two rows of equally spaced electric charges, one row positive and the other row negative (Figure 8-2). A single positive test-charge is located a distance d from these two rows. The force between the test charge and each of the charges in the wire is given by Coulomb's Law. However, since there are an equal number of positive and negative charges in line, and since the magnitude of the positive and negative charges are exactly the same (a very important fact), all of the electrostatic forces cancel out and there is no force on the test charge.

But now suppose that the rows of positive and negative charges are set into motion. (To simplify the description we assume that they go with equal speeds in opposite directions, but the final result does not depend on this assumption.) If the test charge is at rest in the laboratory, it still

Electromagnetism

(a)

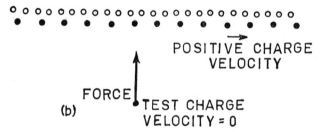

(b)

Figure 8-2. In (a) a row of positve charges moving to the right and a row of negative charges moving to the left are shown as seen in the laboratory frame of reference. There are an equal number of positive and negative charges in each unit length of the wire. A positive test charge a distance d from the wire is moving to the right. In (b) we view the same scene from the frame of reference of the moving test charge. The negative charges are moving to the left faster than the positive charges are moving to the right. Because of the Lorentz contraction, the negative charges, as seen in this reference frame, are closer together than the positive charges. Therefore the test charge sees the wire to be negatively charged and experiences a force of attraction. In the frame of the test charge this is called an electrostatic force. In the laboratory frame it is called a magnetic force.

"feels" equal numbers of positive and negative charges in the imaginary wire containing the rows of moving charges. Therefore it continues to feel no electric force.

However, suppose that the test charge is now moving parallel to the other charges, in the same direction as the positive row of charges. From the reference frame of the moving test charge, the negative charges in the wire are now moving more rapidly than the positive charges. Here is where relativity theory comes in: Due to the Lorentz contraction the more rapidly moving negative charges appear to be located more closely together than the positive charges. (Think of the distance between adjacent charges as a measuring rod that contracts when in motion.)

Now here comes an important question: If the distance between charges changes because of the motion, and if the mass changes, then how about the quantity of electric charge? Does that change or remain constant? The answer is that—as we shall presently see—the amount of charge on a single particle is found to be a constant. As the relativists say, the quantity of charge on a single object is invariant with respect to its velocity. This fact produces an important result: Since the negative charges in the moving line are closer together than the positive, and since the quantity of charge on each particle has not changed, it follows that each unit of length along the wire contains a greater amount of negative charge than positive—from the viewpoint of the moving test charge.

This is a result that is completely unexpected from the classical point of view. It is an entirely relativistic effect: The moving test charge sees the wire to have a negative electric charge, and since the test charge is positive, it feels an attraction pulling it toward the wire. If the test charge is one of a row of charges in a second wire, then the second wire is attracted toward the first. As we look at this occurrence from the outside we don't attribute it to an electric force. We say that this is the result of a magnetic force.

But, as we have shown, this so-called magnetic force is nothing but our familiar electric force that has become unbalanced due to the way in which dimensions become changed in a moving frame of reference. Here, in fact, is a relativistic

Electromagnetism

effect that is important even with charges moving at quite small velocities, because there are so many electrons in a wire that this effect is very easily noticed. This is an exception to the rule, so often quoted, that relativistic effects are important only when we are dealing with objects moving at speeds approaching the speed of light.

If we proceed to apply the Principle of Relativity to calculate the magnitude of the magnetic force between the two current-carrying wires,[2] we find that we obtain the equation

$$F_m = \frac{2I_1I_2(\text{esu})}{dc^2}$$

Notice that the current in this formula is given in electrostatic units, because we started out with the Coulomb Force Law, in which the charges are in esu. Previously we said that if we measure the current in electromagnetic units, the same amount of force is given by the formula

$$F_m = \frac{2I_1I_2(\text{emu})}{d}$$

Comparing these two equations, we see that

$$I(\text{emu}) = I(\text{esu})/c$$

or $$I(\text{esu}) = c\, I(\text{emu})$$

This tells us that to obtain a given force we need exactly c times as many esu charges as emu, which means that the emu of charge is c times bigger than the esu.

In the beginning, this was little more than a mysterious experimental result, but now we have arrived at a new interpretation. The factor c in the equation comes from the relativity formulas used in the derivation of the equation, so there is no question about the fact that c is exactly equal to the speed of light (recall that the relativity formulas all involve factors such as $\sqrt{1 - v^2/c^2}$).

What the new logic shows is that 1. if all positive and negative charges are exactly equal in magnitude, and 2. if this charge magnitude is independent of the motion of the particle, and 3. if the Coulomb Law of Force (the inverse-square law) holds true for charges at rest, and 4. if the relativistic principles are valid, then the Biot-Savart Law of Force between two currents is correct, and the ratio of the emu to the esu of charge must be equal to the speed of light.

This theoretical argument points our way toward deciding on the fundamental experiments of electromagnetism. The Biot-Savart Law is no longer fundamental, but is derived from more fundamental ideas. Similarly, it is no longer necessary to worry about the emu/esu ratio. Instead, we start with relativity as an established structure. Then, what we must examine for the rest of this chapter are the experiments that verify equality of charge, the principle of charge invariance, and Coulomb's Law. From these experimental results, plus the Principle of Relativity, stem a large fraction of everything we know in the huge subject of electromagnetism.

2. Equality of Electric Charge and Conservation of Charge

One of the great unsolved mysteries of physics is the fact that the amount of electric charge carried by an electron appears to be exactly the same as the amount of charge carried by a proton, even though their masses are quite different, and even though they belong to entirely different classes of elementary particles. There is no theoretical explanation for this fact, and all we can do is say that electric charge always comes in the same kind of unit size, regardless of the type of particle it is sitting on. This unsatisfying explanation has given rise to theories (such as the current quark model of elementary particles) that picture all of the various types of particles as being composed of combinations of a very few "most elementary" particles.

The Millikan oil drop experiment was the first experiment to definitely establish the quantization of electric charge and to measure the quantity of this fundamental charge unit. In that famous experiment the motion of a tiny oil drop in an electric field was measured. If the drop had a few more electrons than normal or a few less electrons than normal, the drop was propelled through the air by means of the electric force, and the amount of charge could be deduced from the measured speed of the droplet. Since it did not seem to make any difference whether the charge was positive or negative, it was concluded that electrons and protons both had the same amount of charge.

Of course, the same conclusion could be inferred from the common observation that ordinary matter is neutral in charge, and since this matter consists of equal numbers of electrons and protons, the charge on each particle must be the same. However, conceivably there might be a very tiny unbalance of charge that could not be observed casually, but could be detected by sensitive experiments. Furthermore, the astrophysicists R. A. Lyttleton and Hermann Bondi have suggested that if the charges on electrons and protons differed by only one part in 10^{18}, the unbalanced electrostatic forces would cause the universe to expand.[3] (This suggestion was made during the heyday of the steady-state theory of the universe, which required some mechanism other than a big bang to explain why the universe was expanding.)

For these reasons a number of experiments have been performed during recent decades to try to pin the charge difference down to as small a value as possible. As in any null experiment, it is impossible to say that the difference between the electron and proton charge is absolutely zero, but one can put an upper limit on the amount of difference.

A modern-dress version of the Millikan experiment was performed by R. W. Stover, T. I. Moran, and J. W. Trischka at Syracuse University.[4] Instead of using an oil drop, they employed a small iron sphere (0.1 mm in diameter) suspended by means of a magnetic field in the air between two parallel metal plates 3 mm apart. Any unbalanced electric charge on the iron sphere would cause it to move in a horizontal direction when a high voltage was put across the parallel plates, and this motion could be detected by the shadow of the sphere on a photocell.

The results of this experiment were that the difference in charge between the electron and proton was found to be about 1×10^{-19} that of the electron charge. A by-product of this experiment was the observation that the light illuminating the iron sphere did not carry enough electric charge to be noticed. Specifically, the charge carried by each photon of light was found to be less than 10^{-16} that of the electron charge. Normally we take it for granted that electromagnetic radiation is uncharged, but it is reassuring to have experiments that verify this assumption.

Another method for measuring charge equality has been developed to an even greater level of precision than the previous experiment. This method is based on the simple idea of mounting a tank of a pressurized gas on insulators within a shielded compartment (Figure 8-3). The gas is allowed to

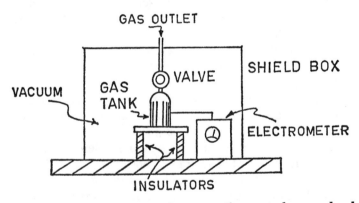

Figure 8-3. An experiment to determine if a neutral gas molecule carries a net electric charge. A gas tank is mounted on insulators inside a vacuum box. As the gas is allowed to leak out through the valve, an electrometer determines if there is any change in potential of the isolated tank, due to charges being carried away by the gas molecules.

leak out slowly, and the electrical potential of the tank is measured with a sensitive electrometer. The theory of the experiment is based on the fact that the molecules of the gas

Electromagnetism

are made of electrons, protons, and neutrons. If the charges within each molecule do not add up to zero, each molecule will carry a little charge out of the tank as the gas leaks away. If that happens, the electrical potential of the tank with respect to ground will change, and this rate of change will depend on how fast the gas leaks out of the tank. Very small changes of potential can be measured with highly sensitive electrometers, so that extremely minute charges can be detected with such an experiment.

One such measurement was made by J. G. King, at MIT,[5] while A. M. Hillas and T. E. Cranshaw published results of similar experiments performed at the Atomic Energy Research Establishment, at Harwell, England.[6] King found the net charge on the hydrogen molecule to be less than $(7 \pm 2.5) \times 10^{-20}$ times the charge on the electron.

Hillas and Cranshaw compared the charges on argon atoms with the amount of charge on nitrogen molecules. Since an argon atom has eighteen protons, eighteen electrons, and twenty-two neutrons, while a nitrogen molecule has fourteen protons, fourteen electrons, and fourteen neutrons, such an experiment permits an estimate of the amount of charge on the neutron, if any. The experimenters found the charge on the argon atom to be less than 8×10^{-20} of the electron charge, while the nitrogen molecule had less than 12×10^{-20} electron charge. From these numbers they deduced that the electron-proton charge difference was less than $(1 \pm 4) \times 10^{-20}$ electron charge, while the charge on the neutron was less than 4×10^{-20} electron charge. This measurement of the electron-proton charge difference is considered the most accurate to date.

The neutron charge measurement is quite an important one for fundamental reasons. We normally take it for granted that a neutron is electrically neutral, but as in the case of the photon, it is important to have experiments verify this to a high order of precision. The reason for this is tied up with one of the most fundamental laws of physics, conservation of electric charge. In elementary-particle physics, this law is just as important as conservation of energy and momentum. The law states simply that in any kind of reaction in a closed system the total number of electric charges remains constant, taking positive and negative charges into account.

For example, in the beta decay of a radioactive nucleus, a neutron changes into a proton while an electron and an antineutrino is emitted:

$$n \to p + e + \bar{\nu}$$
charge: $\quad 0 \to 1 - 1 + 0$

If the neutron has exactly zero charge (and also the antineutrino), and if the electron and proton charges are exactly equal, we see that the total charge on the left side of the equation is zero, while the charge on the right side adds up to zero. The electric charge is conserved in this reaction. Thus the precise measurements of charge equality and the neutrality of the neutron are seen to be the experimental bases on which we base our belief in the Law of Conservation of Charge.

3. *Invariance of Electric Charge*

Two of the fundamental properties of any object are its mass and its electric charge. Both of these properties were always considered to be constant, invariant, "inside" quantities that did not depend on any external factors such as the location of the object, its velocity, the presence of electric or magnetic fields, and so on. Yet, with the advent of relativity a curious split took place. The mass of an object—relating to its inertia and its gravitational interactions—was found to be not really a constant, but depended on its velocity. The intrinsic, unvarying property was found to be the rest-mass. The quantity of electric charge, on the other hand, did not appear to depend on the velocity of the object at all. That is, an electron accelerated by the constant electric field between two parallel charged plates experiences the same force regardless of how fast it is going. (The motion here is assumed

Electromagnetism

to be parallel to the direction of the electric field; otherwise the particle does feel a force altered by relativity.)

Another indication that the amount of charge on a particle is independent of its motion is the fact that the charges on an atomic nucleus are all exact multiples of the proton charge. It is this fact that allows an atom with equal numbers of electrons and protons to be electrically neutral. For example, in the experiments described in the previous section, the electric charges on various kinds of atoms were found to be zero to within one part out of 10^{20}. In the experiment of J. G. King, both hydrogen and helium were tested for charge neutrality. The hydrogen nucleus consists of a single proton, moving with the relatively low thermal speed of the hydrogen molecule. Inside the helium nucleus, however, the protons are whirling rapidly about within the nuclear potential well, with energies in the million electron-volt range. This makes no difference as far as the charge is concerned.

We thus have the remarkable situation where the mass of an atomic nucleus depends on the energy state of the particles inside, but its electric charge does not. If the charge as well as the mass depended on the energy of the protons, it would be extremely difficult to measure masses of isotopes by means of a mass spectrograph, for in all of the calculations it is assumed that the fundamental electric charge is constant. The same is true in all calculations involving atomic spectra, where it is assumed that "e" stands for the charge on the electron, and is the same number regardless of the orbit in which the electron happens to be located.

4. Coulomb's Law

The fundamental law of electrostatic force is Coulomb's Law, with which we started this chapter. This law states that the amount of electric force between two charged objects varies according to the product of the two charges, and inversely as the square of the distance between them. The first accurate measurements of this inverse-square law were made in 1785 by Charles Augustin Coulomb, using his recently invented torsion balance. Coulomb's technique involved a direct measurement of the electrostatic force between two charged spheres located a known distance apart. The method was based on the observation that the twisting of a fiber was directly proportional to the applied torque (Figure 8-4).

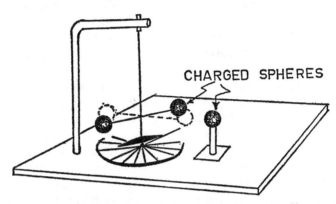

Figure 8-4. Coulomb's torsion balance. A thin quartz fiber is suspended from a support, while a cross-arm is fastened to the bottom of the fiber. A twisting force applied to the cross-arm is detected by the motion of the pointer around a circular scale.

Such a method was sufficient to verify the guess that the force varies inversely as the square of the distance, but its precision was limited by the distance measurement as well as by the force measurement. A more sensitive method invented by Henry Cavendish and later modified by James Clerk Maxwell provided a means of verifying the inverse-square law without measuring any distance at all. Only the detection of a small electric charge was required, and as we have seen such charge measurements are extremely sensitive.

The method is based on a theorem of Johann Gauss, which states that if a hollow conducting box of any shape is electrically charged, and if the electric field strength due to that charge obeys the inverse-square law, there will be no electric field inside that hollow box. That is, there will be no force acting on a test charge placed anywhere inside the enclosure.

Electromagnetism

Any isolated body inside that box will show no tendency to become charged, no matter how much charge is put on the outside of the box. (The reason for this behavior is that any charges put on the hollow box will spread out over the surface of the box due to their mutual repulsion. Any test charge on the inside of the box will feel equal forces pushing it in all directions, so the net result is as though there is no force field inside the box at all.) Gauss' Law is the principle on which electrical shielding operates. Any device inside a conducting box will be isolated from any kind of electric fields on the outside of the box, assuming that the box is a perfect conductor. The essential point here is that all of this behavior depends on the exact validity of the inverse-square law of electric field strength.

The experiment devised by Maxwell involved two insulated metal globes, one inside the other. The outer globe had a small hole in it, so that the inner one could be tested for electric charge by inserting an electrode from an electrometer. The hole was closed by a lid that shorted the two shells together. The outer shell was then charged to a high potential. Following this the lid and connector were removed by a silk thread. According to the theory, if the inverse-square law were true, all of the electric charge should have collected on the outer shell, and none should have remained on the inner globe. Conversely, if the force between two equal charges Q followed the law

$$F = \frac{Q^2}{r^{(2 \pm q)}}$$

then the inner sphere should retain a certain amount of charge and should have a measurable potential from which the quantity q could be found. (The number q simply indicates how much the force law deviates from a perfect inverse-square law. We normally assume q to be zero.) By measuring the potential of the inner sphere, using the gold-leaf electrometers available at that time, Maxwell found q to be less than 1/21,600, a result that was still quoted as definitive up to 1936.

In 1936 S. J. Plimpton and W. E. Lawton, of the Worcester Polytechnic Institute, used some of the refinements of modern electronics to extend the precision of the experiment by several orders of magnitude.[7] The new experiment utilized a metal sphere five feet in diameter that was charged to a po-

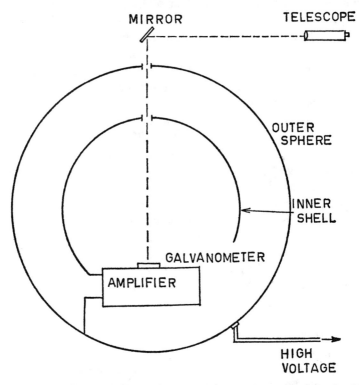

Figure 8-5. Plimpton and Lawton's apparatus for testing Coulomb's Law. The outer sphere (five feet in diameter) is connected to an alternating high-voltage source. The potential of the inner shell is measured by an amplifier and galvanometer. The galvanometer needle is read from outside the sphere through a small window. In the actual experiment, the window glass was filled with salt water so as not to interrupt the conducting sphere.

Electromagnetism

tential of three thousand volts from a high-voltage rectifier (Figure 8-5). At the same time the potential of the inner sphere was measured by a sensitive electrometer amplifier and galvanometer capable of detecting a potential change of ½ microvolt. The outer sphere was charged alternately positive and negative by a motorized switch arrangement at a rate of two cycles per second. There are several reasons for this alternation of potential; an AC amplifier can be used, avoiding the slow drifting that tends to plague DC amplifiers. In addition, the amplifier can be tuned so as to be sensitive only to the frequency range in the neighborhood of two cycles per second. In this way the signal-to-noise ratio is improved and the sensitivity of the system is increased.

By this method the exponent in Coulomb's Law was found to be equal to the integer 2, to within one part in 10^9. That is, the quantity q was found to be less than 10^{-9}. We notice that with this degree of precision Coulomb's Law is found to differ from the exact law of gravitation, where the modifications due to relativity cause the law to deviate from inverse-square.

Since 1936 many improvements in electronics—notably the development of more sensitive electrometer amplifiers, and the use of lock-in amplifiers to extract very small repeating signals out of random noise—have made it possible to repeat this experiment and to squeeze a few more decimal places out of it.*

* A measurement reported by a group at Princeton University used an apparatus with five concentric spheres and a lock-in amplifier with a sensitivity of 2×10^{-10} volt. The upper limit found for the value of q was 1.3×10^{-13}. (D. F. Bartlett, P. E. Goldhagen, and E. A. Phillips, *Phys. Rev. D*, 2, 483 (1970).)
An even more recent measurement performed at Wesleyan University also used five concentric shells. Applied to the two outer shells was 10 kilovolts of rf power at a frequency of 4 MHz. A signal of less than 10^{-12} volts could have been detected between the two inner shells. In this experiment the upper limit found for the value of q was 2.7×10^{-16}, which makes it the most precise verification of Coulomb's law to date. (E. R. Williams, J. E. Faller, and H. A. Hill, *Phys. Rev. Letters* 26, 721 (1971).)

In 1968 a completely different method of obtaining the same result was reported by A. S. Goldhaber and M. M. Nieto.[8] The method was originally suggested by Erwin Schroedinger in 1943, but the required measurements were not available for many years. The method is based on the idea that any deviation of the electrostatic field from Coulomb's Law would also result in a change in the form of the magnetic field produced by moving electric charges. In particular it would cause certain minute changes in the earth's own magnetic field. The earth's field (to a first approximation) is a dipole field—that is, the lines of force are shaped as if there were a bar magnet located at the center of the earth. A deviation from the inverse-square law would produce a change in the magnetic field as if some enormous coil outside the earth were superimposing a constant external magnetic field on the earth's dipole field.

Since the earth's field is under constant survey by geoscientists using both earthbound and satellite instruments, it was possible to analyze the data to look for any possible anomalies —any external fields of unknown origin. As a result of this analysis it was possible to say that any such unknown external field, if existing, must be less than 4/1000 gauss in magnitude. (Earth's field is about ½ gauss at the surface.)

In order to understand the deeper significance of these Coulomb Law measurements, we must at least scratch the surface of the science of quantum electrodynamics. We recall that ordinary quantum mechanics explains the structure of atoms and their nuclei by imagining the electrons and protons and other particles to be wave-type phenomena. With this model as a beginning, Schroedinger and many others were able to explain why electrons in an atom exist only in certain energy levels, give off certain spectrum lines, and so on. This early form of quantum mechanics still uses the classical idea of the inverse-square law of force between electrons and protons. No effort is made to give a deeper explanation of the origin or structure of the electric field.

However, quantum electrodynamics represents a deeper

Electromagnetism

analysis of reality, in which the electric field itself is imagined to be the end result of the back-and-forth exchange of photons between electrically charged particles. There is a continual emission and absorption of "virtual photons" of electromagnetic energy by each of the particles involved—the photons are batted back and forth like tennis balls, and by the mathematical analysis of this process a complete explanation of the electromagnetic interaction has been derived.

One of the essential features of this picture is the postulate that a photon has zero rest-mass. The only mass a photon possesses is due to its kinetic energy. A direct result of this assumption is the inverse-square law of electric force. An additional result is the inverse-square law of electromagnetic radiation—the intensity of the radiation emitted from any kind of source must decrease according to the square of the distance from the source. If the photon had a finite rest-mass, rather than zero rest-mass, the inverse-square law would not hold. The strength of the force would decrease faster than $1/r^2$.

It seems as though we are relying on a very complicated trick to explain a simple inverse-square law by this juggling feat of photon exchange. However, there are certain fine features of atomic spectra that can only be explained by this kind of law. In addition, there was a very dramatic vindication of this general type of theory when Hideki Yukawa in 1935 showed how the nuclear force binding neutrons and protons together could be explained by postulating a quantum of energy being exchanged back and forth by the nuclear particles. Since the nuclear force is a short-range force (it decreases with distance much more rapidly than an inverse-square force) it was necessary to propose that the particle exchanged had a mass two or three hundred times greater than an electron. In later years the pi-meson was observed in cosmic ray showers, answering to the description of the entity needed to satisfy the strange new explanation of the nuclear force.

In like manner, a test of the inverse-square law is also a

test of the rest-mass of the photon. A deviation from inverse-square law means that the photon mass is greater than zero. In terms of this analysis, the earlier experiment of Plimpton and Lawton implied that the photon mass was less than 3.4×10^{-44} gram, or 3.7×10^{-17} electron mass. The newer observations, using satellite measurements of the earth's magnetic field, gave an upper limit on the photon mass of 4.0×10^{-48} gram, or 4.4×10^{-21} electron mass. The new figure is seen to be ten thousand times better than the older one. Here we have a prime example of how experiments performed in space have made a profound contribution to one of our most fundamental areas of knowledge.

As a result of the fact that the electric force between charges follows the inverse-square law, there is the further consequence, as we have already remarked, that the radiation of electromagnetic waves also follows an inverse-square law. This can be shown from Maxwell's equations, a rather complex mathematical development carried out in all the standard textbooks on electromagnetic theory. It can also be shown more simply by thinking of electromagnetic radiation as a flux of photons emitted from a source and spreading out equally in all directions. Each photon carries with it a certain amount of energy, and if conservation of energy holds true, the energy of each photon must remain constant as it flies through space. Thus, if the number of photons going away from the source does not change, the total amount of energy spreading out through space remains a constant.

The effects produced by this radiation when it reaches its destination depend on its intensity—the amount of energy falling on each square centimeter of target surface. Suppose we let I represent this radiation intensity, while E stands for the total amount of radiation energy passing through a sphere of radius R. The surface area of the sphere is $4\pi R^2$, and since the total energy is the intensity multiplied by the area, we can write

$$E = I \cdot 4\pi R^2$$

Electromagnetism

If the total amount of energy remains constant as it travels away from the source, then at any distance R the intensity will be inversely proportional to the square of the distance:

$$I = \frac{E}{4\pi R^2}$$

We see that the inverse-square law for radiation follows directly from conservation of energy. In terms of the photon theory we discussed above it means that photons are neither created nor lost as they fly away from the source. For this to be true, the theory requires that these photons have zero rest-mass. If the photons have a rest-mass greater than zero, the intensity of the radiation diminishes even more rapidly than predicted by the inverse-square law.

What happens to conservation of energy if the intensity does not follow an inverse-square law? The photon theory explains this by saying that not all the photons get all the way out to very great distances from the source of emission —some of them are reabsorbed before they escape completely. The point of the argument is that when *any kind* of energy-carrying radiation is broadcast from a source, its intensity must decrease at least as rapidly as the inverse-square law. It cannot decrease less rapidly. This conclusion is related to the fact that space has three dimensions, so that as the energy spreads out in all directions it must pass through greater and greater two-dimensional areas, thus becoming diluted in proportion to the area.

I dwell on this point because it has the most important practical consequences, as we will see in the next chapter. Any kind of signal or message broadcast in all directions must travel by some kind of radiation and so must get weaker as it goes along. An extreme example of this kind of behavior is observed when our radio astronomers send a radar signal out toward Venus and then detect the signal reflected back to earth from Venus in order to determine the nature of Venus' surface. The inverse-square law operates twice in this case, because first the signal is radiated from Earth, and then

Discovering the Natural Laws

reflected from the distant planet. As a result, the radio telescope may transmit millions of watts of power, and then receive back less than a millionth of a millionth of a watt, strictly in accordance with the inverse-square law.

CHAPTER 9

What Is Forbidden?

1. *Events allowed and forbidden*

A point of view that has arisen among many of the scientists working at the frontiers of knowledge is the idea that *anything will happen as long as it is not expressly forbidden by a law of nature.* This sounds almost trivial at first glance, and yet it is a complete about-face from the guideline so prevalent a century ago—the rule that if a proposed action does not conform to a set of preconceived ideas, then it is impossible.

The emphasis of the new idea is optimism—and rebellion against the preconceptions of past generations. It is a fruitful point of view because it guides the scientist toward making new discoveries by imagining every event that might occur under a given set of conditions. If no known law forbids these events, then they ought to be taking place somewhere. If you look for them and you can't find them, then perhaps this is an indication that there is a law of nature that forbids this type of happening, and now you are on the path toward discovering a new law of nature.

This type of philosophy stems from the way atomic and nuclear reactions are governed by "selection rules." When a number of ingredients are combined in a test tube, in a nuclear reactor, or in a particle accelerator, it is easy to imagine an extremely large number of reactions that might conceivably happen, but in practice only those few reactions take place that are permitted by the laws of nature.

Discovering the Natural Laws

For example, if a container of hydrogen gas is bombarded with electrons, one of the expected reactions might be a direct collision between the bombarding electrons and the protons that make up the hydrogen nuclei. Since the electrons and protons attract each other, it would appear possible for the two particles to combine and form a neutron. However, if you compare the energy (or mass) of the neutron with that of the electron plus proton, you realize that the electron would have to possess at least 780,000 electron-volts of energy in order to join with the proton to make a neutron. This is required by conservation of energy. But that is not enough. The electron has a certain amount of spin angular momentum —½ unit in the system used by atomic physicists. The proton also has ½ unit of angular momentum, and so does the neutron. Now angular momentum must be conserved, but there is no way to put the ½ unit of the electron together with the ½ unit of the proton to make the ½ unit of the neutron's angular momentum. In order for the reaction to be possible, it would be necessary to have a neutrino come along at just the right moment so that its bit of spin is added to the system, and angular momentum is conserved. The chances for both the electron and neutrino to hit the proton at exactly the same time are very small indeed, and so we would call this kind of reaction highly improbable, but not impossible. At least our knowledge of the fundamental laws has enabled us to narrow down all the vast possibilities and identify the conditions that must be met for this reaction to happen.

The physicist looks at a number of objects making up a system and he sees it existing in a state of some kind. This system may be nothing more than a single atom, or it may be a solar system, or it may be a group of people—a social system. Whatever the system, either it stays in the original state or it changes to another state, depending on the kind of forces acting on it. The change from one state to another is governed by the selection rules, laws such as conservation of energy, momentum, angular momentum, electric charge, parity, and so on. If the selection rules allow it, the change of state takes

What Is Forbidden?

place. If the selection rules do not allow it, we say that the transition from one state to another is forbidden.

For example, if I am standing on the ground, the earth and I represent a system in its state of minimum energy. Put me a hundred feet in the air, and the system is in another state with a somewhat greater amount of potential energy. Now if I happen to be unsupported, there will be a rapid transition from the higher state to the lower state. It requires no outside force to accomplish this transition—the inner workings of the system always allow it to go from a state of higher to a state of lower potential energy, and in this case the force causing the transition is the gravitational force.

On the other hand, if I begin on the ground and ask how I can get up to the hundred-foot level, then I must give careful consideration to the Law of Conservation of Energy. I know that the machinery in my body will not supply enough energy to make the transition in one jump. If I have a ladder I can go up one rung at a time, acquiring the required energy in small steps. If I have a helicopter I can use the energy stored within the fuel to aid my ascent.

The point is that the hundred-foot level is not absolutely forbidden. It is merely forbidden under the restricted condition that I am not allowed any outside sources of energy or any mechanical climbing aides. Once more, knowledge of the laws of physics has allowed us to specify what must be done in order to reach a desired goal—to go from one state to another.

On the other hand, consider a state that consists of me traveling at a speed of one million miles per second. Such a state is absolutely forbidden by the behavior of matter as observed in the laboratory or elsewhere in the universe. There is no way for any object made of real matter to reach a speed greater than the speed of light. This dogmatic-sounding statement must stand until we find a kind of matter that behaves differently than all the objects we have ever observed.

There is another kind of forbiddenness that is not absolute —a type of event that can happen, but somehow never (or

Discovering the Natural Laws

hardly ever) does. Think of a container filled with water and ink in equal quantities. If you start out with the water all on one side of the container and the ink all on the other side you certainly expect that during the course of time all of the molecules will become mixed up and the two liquids will be uniformly intermingled. You have no expectation that once the liquids are mixed they will proceed to unmix themselves. If you waited a million years you would not find the water once more at one side of the container and the ink at the other side.

Now there is no fundamental law that says it can't happen. As long as all the conservation laws are satisfied there is nothing to forbid the unmixing on an atomic level. The only reason it doesn't happen is that it is a highly unlikely process, leading to a very improbable state. There are many, many other states that are equally allowed and, since there are more of them, there is more chance that the system will reach one of these more common states—just as there is more chance of drawing a pair than a royal flush in poker. And it so happens that all of these more numerous states are those in which the two fluids are more or less uniformly mixed. We do have a law of entropy that describes how systems tend to approach a state of greater disorder as time goes on. This is a law that does not deal with individual particles, but concerns itself with the average behavior of large groups of objects.

You notice that in all this discussion of the kinds of events that can happen, I have not once used the words *possible* or *impossible*. The reason is that these words have too many confusing and derogatory meanings, and are too prone to be used unthinkingly and with prejudice. There are at least three common meanings of the word *impossible:* mathematically impossible, physically impossible, and psychologically impossible. (I can't do it, I don't know how to do it, I'm afraid to do it, I don't have the means to do it, etc.)

A mathematician will say that a certain operation is impossible within the set of rules governing that particular type of

What Is Forbidden?

mathematics. For example, in Euclidean geometry it is absolutely impossible to draw more than one line through a point parallel to a given line. Change the rules of the game to non-Euclidean geometry and the impossibility turns into a possibility. Of course, games themselves are filled with such impossibilities. In chess it is impossible for a knight to advance three rows in a single move. You may do so if you like, but then you are inventing a different game. The features that characterize games and mathematics are that the rules are made by men; they are not inevitable.

Often in mathematics certain rules do appear to be necessary because we are accustomed to using them in connection with physical objects. We know that $2 + 2 = 4$ because that is the way apples and blocks of wood behave, and so the mathematical rules were made to correspond. Rules that apply to physics appear inevitable because these rules have been made up according to the way things work in nature. When we say it is impossible to build a perpetual motion machine, what we mean is that we have in our minds a concept called conservation of energy, and if that concept really describes how things work in this particular situation, then it is physically impossible to build a perpetual motion machine.

In this sense the word *impossible* means the same as the idea of *forbidden* that we just discussed. A great deal of physics is devoted to finding out what happenings in nature are allowed and what are forbidden, or—most important—under what conditions an event is allowed to happen. This is really the function of the most fundamental laws of nature.

Too often we get the idea that these laws deal only with things happening in the laboratory, but if these ideas have any validity at all, they must extend to every area of human experience. The laws of physics have something to say about every conjecture ever proposed, whether in biology, psychology, parapsychology, or philosophy. To illustrate how this works, suppose we take a critical look at a number of ideas

Discovering the Natural Laws

that have been very popular in recent years and try to decide what the laws of physics have to say about them.

2. Physics and Science Fiction

Science fiction is a branch of literature that has its roots in Greek mythology, but that expanded into its present vigor during the first decades of the twentieth century. That was a time when modern physics was just beginning to understand the nature of the atom, of space, and of time. It was easy in that era of semiknowledge to invent grand new concepts, new fictional machines for the technological world of the future. Space travel, faster-than-light travel, teleportation, time travel, telepathy, atomic energy, antigravity, thinking robots, artificial life, control of evolution, and so on. Many of these ideas were invented around the turn of the century, and all of these are still an integral part of the fictional future world that seems to exist with its own independent life in the literature of science fiction, regardless of what our increased knowledge of science has to say about it.

A number of the predictions of science fiction have come true. Space travel and atomic energy are the most prominent examples. In spite of what a number of pessimists claimed about the impossibility of space travel, there never was a law of nature forbidding it. It was simply a technical problem of getting enough concentrated energy to achieve escape velocity, plus the complex control system necessary for guidance. The complexity of the entire operation is the only part of the story that was never quite realized by the early science fiction writers. When we look back at the stories in which our hero built his own spaceship with the help of a few cronies, we find that truth is not only stranger than fiction, but it is usually more complicated.

The same thing can be said for atomic energy. In the case of fission power nature was kind to us by setting up conditions that just barely made atomic energy practical as well as possible. We must always remember that human beings cannot push nature around. We cannot make things do what is for-

What Is Forbidden?

bidden. All we can do is to arrange objects so that, when they do what they have to do, the results are useful for our purposes.

So it is very fortunate for mankind that during the fission of uranium, about 2.5 neutrons are emitted from each fissioning nucleus, on the average. If this number of neutrons happened to be less than *one*, a fission chain reaction could not take place. If the number were less than *two*, then even if we made chain reactions, we could not increase our supply of fuel by breeding U^{238} into plutonium, and we would therefore be limited mainly to the smaller amount of U^{235} in the world. On the other hand, if the number of neutrons were much greater than three, it would be too easy to make chain reactions go, with possibly disastrous results. Here, then, we were lucky in a practical sense: Fission was not only allowed, but it was allowed in a useful way.

Fusion power is another story. We know how to fuse deuterium and tritium nuclei so as to give off the enormous energy of the hydrogen bomb. We also know in theory what conditions must be set up to accomplish thermonuclear fusion in a slow and controlled manner for generating useful power. After twenty years of research, scientists are still trying to set up those conditions. In this area nature has not been cooperative, and the full totality of all the complexities in the problem are just starting to be understood. Yet there is no law that says the job can't be done, and this is the optimistic outlook that must be held by the men working in this field. Here we have an operation that is not forbidden by any known law. In the future such a law may be found. There is the possibility that there is some behavior of extremely hot, ionized gases—such as turbulences created by an unstable state—that absolutely prevents the conditions for controlled fusion to be reached. Clearly, our present knowledge is not enough to make a definite prediction one way or the other. But as long as we don't know it is forbidden, we keep trying.

On the other hand, consider the idea of antigravity. Here is a prime example of a universal wish: the desire to be levitated, to float effortlessly in defiance of the force of gravity, to ease the departure from the earth—basic ingredient of much science fiction. Unfortunately, there is nothing in the behavior of matter that gives us hope that such a dream can be realized. As we have already described, gravitation is a universal force, causing all objects to attract—never repel —each other. Electrons, protons, neutrons, and photons all obey the same law. Since all observed matter is made up of these objects, there is no way of arranging any kinds of matter so that they do anything other than to attract each other and to attract other masses—as far as the force of gravitation is concerned. The gravitational force cannot be shielded or otherwise altered because there is only one kind of mass, in contrast to the two kinds of electric charges.

As a result of these observations, we can make quite a definite statement: *If* the only kind of matter available in this universe obeys the law of gravitational attraction described in Chapter 4, *then* there is no way to construct an antigravity device.

Notice that there is no dogmatic reliance on theoretical assumptions. There is the recognition that all the matter we observe behaves in a certain way, and if this is the only kind of matter there is then we are never going to see any other kind of behavior.

The same kind of argument holds for the idea of traveling faster than light. Our experiments assure us that any object with a finite rest-mass cannot even get up to the speed of light, let alone go past it. To do so would require an infinite amount of energy. Again, this is the way ordinary matter behaves—this is the way we observe it. It's not just a theoretical idea. As for tricks such as inertialessness, space warps, and the other devices beloved of science fiction writers, these are really little more than magical incantations that allow the author to ignore the way the universe is, so that he can write about it as he would like it to be.

Here, of course, I get into the difficult position of sounding dogmatic if I say flatly that space warps are not allowed. On

What Is Forbidden?

the other hand, there is nothing about the observed behavior of all the known particles (together with the four kinds of interactions) that gives us a hint that they can be arranged in such a way as to produce a space warp capable of allowing a vehicle to slip through from one part of the universe to another. (Although a possible exception to this statement might be the "black holes" in space where stars have undergone gravitational collapse. The suggestion has been made that such holes may connect different regions of real space. It must be kept in mind that this idea is still an untested hypothesis, and that even if it were true there is the probability that a spaceship going through such a hole would be torn apart by the intense gravitational forces.)

Also, I must hedge my bet by noting that I haven't said faster-than-light travel is impossible. I have merely said that it is not allowed by any observed behavior of matter and energy. Similarly for the idea of sending messages by faster-than-light signals of some kind. Until recently, all the relativity experts would have said flatly that this operation is impossible, because no kind of signal or energy can go faster than light. Recently the idea of *tachyons* has been proposed. These are particles that (according to the hypothesis) always travel faster than light, and never travel slower than light. (This feat is made possible by the fact that the rest-mass of such a particle is an imaginary number, and so the actual dynamical mass is a real number.) If it is possible to send signals by means of these particles, we get into a complicated technical argument involving the paradox of signals traveling backward in time, and the journals are indeed filled with arguments on this subject. However, the proof of the pudding is in the eating. The authors of the tachyon hypothesis claim that in the interaction of tachyons with ordinary matter there is produced radiation that allows the shadowy presence of the tachyons to be deduced. Without the possibility of indirect detection of tachyons the entire hypothesis would be empty because it would not predict any observable results. If tachyons are actually observed experimentally, their existence will be proved. So far they have not yet been observed. Whether or not it is possible to send signals with them is a separate question. At any rate, until they are observed, the tachyon hypothesis is a clever exercise in imagining the unimaginable, and in clarifying what Einstein's theory says and what it does not say.

3. *Physics and Life*

The atoms inside our bodies are the same kind of atoms that exist everywhere else. To a modern scientist this statement appears so obvious that it is not worth discussing. Yet, the phrase "organic compound" reminds us that there was a time not so long ago when chemists believed that certain kinds of molecules could be produced only by living creatures possessing a "vital force." We have now progressed to the point where we can produce in the laboratory hundreds of thousands of organic compounds, some existing in nature and others brand new. Not only that, but we are approaching the point where we can synthesize real living matter. (At the time of writing, the first complete gene had been put together in the laboratory.)

The rise of molecular biophysics is based on the assumption that the fundamental laws that govern individual atoms and simple molecules also govern the combination of these atoms into more complicated molecules. Of course, as we go up the scale from simple structures to more complicated ones, new laws are discovered that deal with the organization of the simpler molecules into the more complex structures. However, these new laws do not compete with or contradict the fundamental laws of physics—they simply allow us to visualize more simply and easily the operation of very large systems of molecules. (For example, when we describe the flow of various kinds of gases and liquids, we could start out by writing down the equations of motion for each of the molecules in the fluid, but then we would have the problem of solving millions of individual equations. To make a solvable problem we combine all the individual equations into

What Is Forbidden?

one small set of a few equations that describes how the entire conglomeration of molecules behaves on the average. These are the fluid equations of hydrodynamics.)

The point I am making—a point that I want to emphasize very strongly—is that living beings cannot perform any actions that are forbidden by the fundamental laws of physics. Living creatures can do many things that can't be done by inanimate objects, at least at the present time. They can reproduce, they can talk to each other, they can feel pain, they can write novels and poetry. But they cannot do anything that is forbidden by conservation of energy, momentum, and all the other laws we have been talking about. In other words, living things do not possess any kind of "vital force" or "psychic power" that allows them to engage in actions outside the bounds of natural law.

At this point we plunge into an area fraught with controversy. Books on basic physics don't commonly talk about extrasensory perception (ESP), astrology, psychic phenomena, and the like. Yet, if physics represents the body of laws covering all the things in nature, then it must include the actions going on in the mind.

Thinking processes are made up of an unimaginably complicated sequence of electrical nerve pulses, chemical changes, memory storage by the alteration of molecular patterns, and so on. The biophysicist analyzing what happens in the brain pays no attention to the possibility of mysterious psychic forces coming in from outer space, or existing somehow separately from the workings of the nervous system itself. He works on the basic assumption that everything inside the body is made of molecules interacting by electromagnetic forces (plus the rules of quantum mechanics), that these molecules are made of atoms no different from ordinary atoms, and that these atoms are made of electrons, protons, and neutrons no different from the electrons, protons, and neutrons existing anywhere else.

Physics at the present time is powerless to describe how these atoms are actually organized to make a thinking brain.

Discovering the Natural Laws

That is the ultimate task, and of course when it is accomplished it will be called biophysics, biochemistry, cybernetics, and no doubt a number of as yet undiscovered hybrids naming our efforts to describe organized complexities. But beneath it all is basic physics, describing the ground rules as laid down by nature.

Let us see how some of these rules apply to the phenomenon of telepathy, enthusiastically studied by so many people. Telepathy is the process of sending thought messages directly from the brain of one person to the brain of another person. In physics, whenever we talk about sending messages or signals we require that some physical medium carries energy of some sort from the sender to the receiver. Voice messages are carried by sound waves—oscillations in the air or other medium. Telephone messages are carried by variations in the electric current passing through a wire. Radio signals are carried by oscillations in an electromagnetic field or, if you prefer, by photons sent out by the transmitter and absorbed by the receiver.

We expect, then, that if a telepathic message is received by any person, this message must have been emitted by the brain of another person, and it must have traversed the space between the two people by some physical means. In traveling through this space the signal must travel with some definite speed, it must carry energy from one place to another, and it must obey conservation of energy, if it follows the laws of nature at all.

Why do I say a signal must carry energy? The answer is simple: A thought arising in your brain is the final conscious end result of a process involving the motion of millions of electrons throughout the nervous system. Regardless of all the complications, if a message comes into your brain and starts a thought going, this message has to push some electrons around, and that takes energy.

How can energy be transmitted through empty space? This brings us back with a jolt to the very foundations of physics. We only know of four kinds of forces in nature that

What Is Forbidden?

have been definitely identified, and there are two more types of forces that have been hypothesized, but not yet observed. The four known forces are the four different ways in which elementary particles can influence each other: the two nuclear forces, the gravitational, and the electromagnetic force. The nuclear forces do not extend beyond the boundaries of the nucleus and so are useless for transmitting messages over long distances. (The same applies to the two hypothetical forces mentioned.) The gravitational force is hopelessly weak. The most sensitive instruments ever devised are now being used to detect the gravitational signals produced by the explosions of entire stars. It does not seem reasonable to believe that events taking place inside one brain could be transmitted by gravitational forces to another brain and then detected into a coherent message.

This leaves the electromagnetic force as the only one capable of carrying detectable messages through space. Now we know something about this kind of force. We know that messages sent by electromagnetic waves travel with the speed of light. We also know that messages broadcast in all directions obey the inverse-square law as a consequence of conservation of energy. We can also measure the electromagnetic fields in the vicinity of a thinking brain, and we find that they are so weak that in order to detect them at all the experiment must be done in a carefully shielded room to eliminate all of the other electromagnetic signals in the vicinity, signals originating from house wiring, radio broadcasts, near and distant lightning, the sun, the stars, etc. Is it plausible to believe that such electromagnetic waves could carry the complex kind of information necessary to send a thought into somebody else's mind?

Plausibility, of course, is no argument. We must make the hypothesis that telepathic signals are transmitted by electromagnetic waves, and then proceed to test the hypothesis. If telepathic messages were indeed being sent by electromagnetic waves, they should obey the inverse-square law, and the strength of the received signal should diminish rapidly as the distance between transmitter and receiver is increased. (Would you expect to hear from somebody across the continent with the same ease as from a person in the next room?) Furthermore, the transmission should be completely interrupted by putting the receiver inside a box made of copper screening. These are physical experiments that can be performed.

What is the chance that the telepathic message is transmitted by some kind of force we have never encountered in physics, a kind of wave only generated by a living brain? This is the kind of question that cannot be answered without sounding dogmatic. I can only describe what scientists say about questions like this. I think that what most physicists would say is this: Forces are believed to originate as interactions between elementary particles. One of the following two things would have to be true in order to generate a completely new kind of signal for brain-to-brain communication: Either a brand-new kind of particle exists in the brain and nowhere else, or else the electrons and protons in the brain are put together in such a way that they make a kind of wave that's never been produced anywhere else and cannot be detected by any kind of instrument other than another brain. Both of these hypotheses contradict the idea that the living matter in the brain is made of particles no different from those in any other kind of matter. At the very least, any tests of ESP should come to grips with questions such as these.

It is clear that if we think about the possibility of telepathy while trying to keep in contact with the principles of physics, we immediately get into problems of supreme importance. Suppose it could be established that telepathy is a reality, and suppose it could be shown that telepathic messages are really sent through space by means of a physical force previously unknown. This would be one of the most important scientific discoveries ever made. Because of this we cannot treat the subject lightly. At the same time we must be very careful in accepting the results of experiments performed to verify telepathy or other forms of ESP.

What Is Forbidden?

Such experiments generally involve statistical correlations—for example, the kind of test where one subject looks at a number of cards printed with various symbols while the receiving subject writes down the symbols that come to his mind. If the number of correct guesses is better than what you might expect from sheer chance, this result is considered a positive correlation.

The first thing to say about this kind of experiment is that you cannot prove anything by correlations alone. Correlations can merely *suggest* that two phenomena are related directly to each other. To demonstrate satisfactorily that event A is the cause of event B it is necessary to show a step-by-step chain of physical interactions starting at event A and ending at event B.

For example, if John dreams that something terrible is going to happen to Mary, a thousand miles away, I want to know how that information got into the mind of John. As a scientist I am simply not satisfied with an explanation that "it just happened" without any kind of physical interaction traversing that thousand miles and causing a rearrangement of the molecules in John's brain.

When you are looking for correlations that are just a little bit better than what you might expect from sheer randomness, as in a telepathy experiment, you must be very careful with the handling of the data. It is very easy to fall into the trap of taking a run of luck as a positive result, and discarding the rest of the data, or stopping the experiment when the luck runs out. Stories of individual strange and unlikely occurrences are useless, for in a world of over three billion inhabitants, there is a good chance that things are going to happen by coincidence that you would ordinarily consider extremely improbable. These stories make the headlines and you never hear the others.

Correlation studies are useful when applied properly. The effect of cigarette smoking on human cancer was first noticed by statistical methods. More smokers than nonsmokers got cancer, on the average. At first this correlation was only able to suggest that there was a causal relationship between smoking and cancer, and further detailed work was required to make the evidence convincing. In this case it was not too hard to convince the average scientist, because after all the smoke was in immediate contact with the lungs and it was known that certain chemicals present in cigarette smoke do cause cancer when in contact with tissue.

However, if you try to show a correlation between the thoughts in the mind of subject A and the thoughts in the mind of distant subject B with no detectable message, signal, or other form of information passing between the two individuals, the average scientist is going to resist believing that there is a real causal relationship between A and B. If it is shown that B can be aware of the thoughts in the mind of A with nothing passing between them, this event could completely demolish one of the most important foundation-stones of science: the concept of causality.

While it is fashionable these days to challenge the concept of causality and to make hypotheses about events taking place without it, if you can't show that these things actually happen, then the hypothesis is rather empty. The stakes are too high for any but the most rigorous evidence to be accepted. The laws of physics work too well to be overthrown by wishful thinking.

In all of this analysis I have tried to avoid absolute judgments. I have tried to show how we can find our way through these difficult questions by the use of "if . . . then" type of logic. I have consistently pointed out that *if* a certain hypothesis or law is valid, *then* certain consequences follow. This kind of treatment at least allows us to see what our present view of nature says about ideas such as ESP, and points out what alterations must be made in our basic concepts if ESP is found to be a real thing in nature.

My own prejudices are out in the open. I believe that anything that happens is part of nature. To speak of a supernatural is to create a paradox. I also like to think that anything that happens is explainable in terms of physical

What Is Forbidden?

phenomena. Therefore I think that the miracle of human thought will eventually be explained, and that we will some day approach human-type thought with a man-made brain. As for ESP, I take the same attitude that the Patent Office has toward perpetual-motion devices. Bring me a working model and I'll accept it.

4. A View of the Universe

Clearly, the scientist whose head is filled with visions of electrons, protons, neutrons, photons, mesons, neutrinos, and other little particles, all buzzing around each other and building up fantastic structures of living organisms, has a different picture of the universe than the citizen who has no idea of structure, forces, interactions, and causes. The ideas of the nonscientist about the nature of the world are little different from those held by the average citizen of the Middle Ages.

I have students coming into an elementary physics class who believe that the force of gravity stops above the earth's atmosphere—that's why astronauts in space are weightless— and that things stop moving if you don't keep pushing them. Very typical pre-Galilean notions.

This is not said in a derogatory way, for medieval ways of thinking are perhaps natural ways of thinking. After all, everybody can feel for himself that the earth is at rest, and it takes considerable indoctrination to start believing that we are all traveling eighteen miles per second around the sun. It is easier to believe that our activities are controlled by mysterious psychic forces and by the signs of the Zodiac than to comprehend the more complicated and more difficult concepts of modern psychology.

Furthermore, it is more comfortable and comforting to think of the human psyche as more than just the end result of the workings of the nervous system. To think of a human being as nothing more than a conglomeration of atoms and molecules seems cold, inhuman, and meaningless.

But of course the whole is more than the sum of its parts: it includes organization, information, feedback, awareness of existence, consciousness of likes and dislikes, love and hate, pain and curiosity. The meaning of existence is the meaning that each individual gives to his own existence. Understanding the way this comes about does not detract from the importance of individual feelings, but rather should free the individual from unrealistic and superstitious fears and beliefs.

Seen from this point of view, scientific research becomes more than just a means toward building more advanced gadgets. It provides us with a rational basis for a view of the universe, an outlook on life. The most fundamental studies of all, the effort to understand the elementary particles and their interactions, is the real rock-bottom foundation of all knowledge. For if we know how the basic particles work, then we can set limits on the behavior of larger structures.

For this reason, research in fundamental physics is not just a luxury, an expensive game played just for fun. The picture of the universe that we construct in our minds determines our entire philosophy and ultimate behavior.

CHAPTER 10

Epilogue (1989)

Since this book was written seventeen years ago nothing has been found in nature to contradict its argument. No examples of energy creation or destruction have been observed. Consequently, conservation of energy is on safer ground than ever. Proposed new theories of matter and energy that have arisen during the past few decades—the grand unified field theories and superstring theories—include as an integral part of their structures the several symmetries associated with the various conservation laws. Experimental verification of any of the new theories would serve only to strengthen the fundamental concept that has guided physics during the past century: that everything consists of a relatively small number of particles interacting by means of a very small number of forces. The newer theories would merely reduce the presently known forces to various aspects of a single force. They would add no new forces.

Reducing all the forces to a single interaction, with known symmetry properties, would give us increased confidence in the validity of the conservation laws as they apply to all particles interacting with each other in any possible manner. In addition, the Lorentz symmetry required by all legitimate physical theories would continue to define the geometry of space and time. From this geometry we deduce the consequence that no real matter, energy, or observable messages of any kind can ever travel through space faster than the speed of light.

In spite of the fact that the recent advances in physics do nothing but strengthen the position maintained by the standard model of matter, the period of time since this book's original publication has seen within the public psyche a mounting belief in mysticism,

Discovering the Natural Laws

psychic phenomena, astrology, reincarnation, ESP, parapsychology, UFOs, and all the other topics lumped together under the heading of *paranormal phenomena*. Groups intent on forcing teaching of "creation science" into the schools as an alternative to evolution have continued to exert pressure even after major defeats in the courts. The "new age" philosophy touted by a famous movie star encourages belief in reincarnation and channeling, so that millions may experience the seductive pleasures of mysticism while enriching the bank accounts of innumerable purveyors of magical crystals. Worst of all, books on pseudoscience continue to outnumber legitimate science books in the stores.

In response, organizations such as The Committee for the Scientific Investigation of Claims of the Paranormal (CSICOP) have been initiated and have undergone a healthy expansion.[1] However, they can exert only a tiny influence on the wave of credulity that continues to sweep over the population.

Joining the fray, I have extended this book's discussion in a new book: *A Physicist's Guide to Skepticism*.[2] In this work I describe simple rules which help decide what kinds of actions are possible and what are impossible. The concepts of *laws of permission* and *laws of denial* are introduced to assist in these decisions. Newton's Second Law of Motion and Schroedinger's equation are examples of laws of permission. They help us predict future motion of objects under the influence of known forces. In most cases it is difficult to make accurate predictions with these laws because of the impossibility of solving the equations exactly and, in the case of Schroedinger's equation, because of the uncertainty inherent in the motion of fundamental particles.

On the other hand, predictions made using the laws of denial are exceedingly precise, for we merely use these laws to say that a given action may or may not take place. Conservation of energy is the best-known law of denial. It tells us that no action may take place that changes the amount of energy in a closed system.

Utilizing our knowledge of the laws of denial we can show that claims of psychic phenomena, ESP, astrology, and other paranormal phenomena are in the same class as claims for perpetual-motion machines. The classic science-fiction concepts of anti-gravity, instantaneous communication, faster-than-light travel,

Epilogue (1989)

and time travel must be judged with the same kind of skepticism.

After spending time arguing with supporters of pseudoscientific ideas we begin to perceive a variety of patterns emerging in the responses of the believers. It soon becomes apparent that their arguments are to a large extent based on myths about the nature of science that have become disseminated throughout popular culture. To demythologize these ideas we must strip away a number of illusions fondly held by the public—illusions often spread by scientists and philosophers. The inevitable consequence is fierce resistance. We are speaking of wiping out hundreds of years of attitudes and opinions. Nevertheless, the task must be done.

The myths about science arise from a public inability to fully accept the implications of modern physics. Physicists who are privileged to perform the kind of precise measurements and fundamental experiments described in this book take for granted the fact that nature is essentially lawful. It is the underlying postulate of physics. Members of the general public have little opportunity for hands-on experience with physical measurement. For this reason they are unaware of the sources from which scientific knowledge arises and thus are easy prey for myths, rumors, and slogans.

Some of the common myths are as follows: (1) *nothing is known for sure;* (2) *nothing is impossible;* (3) *everything we believe to be true now is likely to be overthrown in the future;* (4) *advanced civilizations of the future will have the use of forces unknown to us at present;* and (5) *advanced civilizations on other planets will possess great forces unavailable to us on earth.* To discuss these in sequence:

Myth 1: *Nothing is known for sure.*

This is a belief handed down from antiquity. It represents the extreme position of philosophical skepticism. A skeptic inhabiting the outer fringe of skepticism is not even prepared to swear that he/she exists. When Descartes said, "I think, therefore I am," he beat a strategic retreat from just such a position at the far end of the scale of doubting.[3]

Even so, modern scientists are often too modest to claim firm knowledge on any topic. Contemplate a statement by a writer as eminent as Jacob Bronowski: "There is no absolute knowledge.

Discovering the Natural Laws

And those who claim it, whether they are scientists or dogmatists, open the door to tragedy. All information is imperfect."[4]

This statement, of course, must be taken in context. It is the introduction to a chapter in which Bronowski shows that the Heisenberg uncertainty principle causes all measurements to have a finite amount of error. Therefore no measurement can give a result that is known with complete certainty. He then contrasts the humility of scientists, with their awareness of universal uncertainty, with the arrogance of political and religious zealots possessed of absolute certainty regarding the truth of their theories, even though their theories are formed without the benefit of empirical evidence or rational thought processes.

Several comments need to be made to clarify this point:

a. As I have attempted to show in this book, the limit of error on many modern experiments is very small indeed. The fact that conservation of energy has been verified to within one part out of 10^{15} is a specific piece of knowledge. The probable error of the measurement is part of the knowledge. Every experimental physicist is trained to estimate the errors of his measurement and to include the error range when reporting his results. Thus, uncertainty itself is part of pragmatic knowledge. It is not necessary for a piece of knowledge to be *perfect* in order for it to be well-founded.

b. Some bits of knowledge are qualititative, rather than quantitative, and if they are based on proper observation then there can be no doubt as to their validity. For example, we know that the earth is not flat; its surface is not a Euclidean plane surface. It does not matter what the precise shape of the earth is. What we know for sure is that the earth is not flat. Knowing what a thing is *not* is sometimes more certain than knowing what it is.

c. Some knowledge is associated with definitions of terms, and therefore is absolutely certain as long as we agree on the definitions. For example, all electrons repel each other. This is an absolute fact. The skeptic might object: how do we know that there are not a few electrons here and there that attract each other instead of repelling? If they were few enough they might not be noticed. The reason we are sure that these strange electrons do not exist is that an electron is defined in terms of its properties. A particle that behaves differently from electrons cannot be an electron. As a matter of fact there do exist particles that are just like

Epilogue (1989)

electrons, but which display an attraction for electrons instead of a repulsion. These particles are positrons.

d. Some knowledge that appears to be simple and qualitative is actually complex and quantitative. Consider the historical struggle between those who believed the earth is the center of the universe, and those who believed the sun is the center. It seems obvious to our modern eyes, working with our modern instruments, that the sun is the center. But then a modicum of thought forces us to realize that the sun is not exactly at the center, because the earth and other planets are pulling on the sun just as much as the sun is pulling on them. The proper center has to be the point where the center of mass of the solar system is located, and that point is not quite at the center of the sun. But why do we choose the center of mass to be something special? It is merely a convenience to simplify the equations of motion of the planets. What is most important, as was recognized by Newton and those who came after, is the concept that the motions of the planets are determined by forces acting mutually between all of the objects in the solar system. All astronomy and cosmology is founded on this idea. The concept of mutual interactions would appear to be a qualitative idea, and thus a piece of absolute knowledge. The success of this idea, however, is supported by numerous observations and measurements, each of which has a finite error. Therefore the theory that the motion of astronomical objects is determined by their mutual gravitational interaction is a piece of empirical knowledge whose validity is determined by the precision of our measurements. Nevertheless, the cumulative measurements are so good that nobody is about to believe any other theory without good and sufficient reason.

Extending the idea of mutual interaction into the microscopic realm, we arrive at the notion that the behavior of all fundamental particles is determined by the four types of forces acting between pairs of objects. This is the basic paradigm of twentieth-century physics. Every one of the billions of collisions observed in a particle-accelerator experiment confirms this standard model of particle physics. The four forces may be increased to five or six by some experiments currently being run to detect new and very weak components of the gravitational force, but this possibility does not change the basic paradigm.

What would upset the apple cart would be the discovery of

forces acting outside the model of mutual interactions. Such new forces would be, for example, the psychic force necessary to explain ESP and other phenomena of parapsychology. Or the supernatural forces necessary to explain the theories of the creationists. With regard to these forces, which currently are outside the domain of science, the principles of empiricism assure us that it is not necessary to believe in their existence until that existence is demonstrated by experiment.

The humility of scientists with respect to certainty puts them at a disadvantage in the political battle against those who would like to dictate the teaching of pseudoscientific theories in the public schools. The creationist can say: "I am certain that all life was created less than ten thousand years ago. Biologists are uncertain about their theory of evolution, because all scientific knowledge is uncertain. Therefore creationism is worthy of being taught in the schools alongside evolution." The scientist can combat this type of logic only by knowing what parts of his science are known "for sure," knowing what the limits of error on this knowledge are, and knowing how this knowledge was attained.

If we did not know anything for sure, then knowledge would be unstable. Science would be built on shifting quicksand, and today's knowledge could always be abandoned in the future. However, this scenario is denied by the face that scientific knowledge *has* increased exponentially during the past three centuries. The beginning of its rise has coincided with the development of reliable instrumentation that frees us from our reliance on uncertain human sensory input. If knowledge were so uncertain that it disappeared as rapidly as it was gained, then there could not be the steady increase that we observe. It would be like pouring water into a bucket with a large hole in the bottom. The steady increase in our body of knowledge is proof that at least *some* of what we know is known for sure. (By the same token, pseudosciences such as parapsychology are identified by the fact that the amount of knowledge in those fields never increases.)

Is anybody seriously prepared to doubt that the earth is round, that matter is made of atoms, that quantum theory describes atomic structure to a high degree of accuracy? To think that nothing is known for sure is to succumb to mythology. Often it is a symptom of a hidden psychological agenda, an effort to leave

Epilogue (1989)

room for comforting mysticism in a world where pragmatic science is taking over all knowledge.

Myth 2: *Nothing is impossible.*

This statement is a slogan, not a law of nature. It has no scientific significance and counterexamples are easy to find. It falls within the province of exhortations by athletic coaches and inspirational teachers. It has nothing to do with physics, for we have already seen that one of the important functions of physics is to search among all the actions that can be imagined within the real universe and to separate the possible actions from the impossible.

The idea that nothing is impossible follows logically from Myth 1. If nothing is known for sure, then perhaps the laws of physics (which tell us what things are impossible) are not always true.

In addition, the idea that nothing is impossible is a psychological outgrowth of the behavior of infants and small children who want their own way all the time and who acquire the belief that they can do anything they want to do. When they grow up they lash out at authority figures who tell them what they can do and cannot do. "No professor is going to tell me what I can't do!" is the rallying cry. "Don't tell me I can't go faster than light."

But it is nature that determines what can and cannot be done. We know as confidently as we know anything that everything is made of elementary particles, and that these particles interact by means of a few fundamental interactions: electroweak (a combination of electromagnetic and weak interactions), gravitational, and strong nuclear interaction. Conservation of energy has been experimentally verified to a very high degree of accuracy for all of these interactions. Therefore everything that happens in the universe, without exception, must obey the law of conservation of energy.

With the help of this law and other laws of denial we can make exceedingly precise predictions of the things that cannot happen. Those of a mystical bent are fond of pointing out that science cannot understand how the brain operates and cannot predict how human beings are going to behave. They take this as proof that psychic activities take place within the human mind. However, it is illogical to assume that because we cannot as yet understand something, this is sufficient cause to assume mystical or supernatural causes at work. There are many inorganic phenomena in the world (such as the weather) which are too complex to allow exact prediction and total understanding. Nevertheless, no scientist assumes supernatural causes for these phenomena. Similarly, scientists assume, as a matter of course, that the laws of nature apply to living beings as well as to the nonliving universe. Therefore, even though we cannot predict what humans are going to do, we can be very confident when using the conservation laws to predict what things human beings *cannot* do.[5] A human being cannot jump to the moon without the aid of a mechanical device. A human being cannot glow in the dark with the energy of his own metabolism. And a human being cannot send energy instantaneously through empty space from one mind to another for the purpose of conveying a telepathic message.

Myth 3: *Everything we believe to be true now is likely to be overthrown in the future.*

This argument is often raised by people who have heard of Thomas Kuhn's influential book *The Structure of Scientific Revolutions*[6] and understand its contents to some extent. Kuhn's thesis is that scientific revolutions take place when a paradigm shift occurs within the community of scientists. This paradigm shift is a change of mind-set in which everybody learns to look at nature in a different way. The change from the earth-centered to the sun-centered solar system was such a historic paradigm shift. The change from Newtonian to Einsteinian (relativistic) physics was another shift.

When enthusiasts pushing pet ideas such as creationism or faster-than-light travel are backed into a corner, their ultimate defense is the argument that "Paradigms have changed in the past. Therefore they are likely to change in the future." Ergo, scientists are eventually bound to give up evolution in favor of creationism, and greater knowledge will force scientists to give up the relativity paradigm so that spaceships may eventually travel faster than the speed of light to the distant stars.

However, a careful reading of Kuhn's book, together with a modicum of common sense, leads us to realize that the above statement is simply wishful thinking. Nowhere does Kuhn say that every paradigm *must* change in the future. In the past, change of scientific opinion has generally taken us from belief in false

Epilogue (1989)

paradigms to belief in correct paradigms: from belief in a flat earth to a round-earth model, from the caloric concept of heat to the kinetic theory of matter. Once a paradigm has been shown to represent the real world, why should it change again?

In this context the word "paradigm" refers to a *model*. It refers to a set of concepts that attempts to describe the structure of the universe so that we may explain how things work. We have seen, under Myth 1, that the fundamental twentieth-century paradigm is the idea that all events are governed by interactions between pairs of particles. The form of each interaction is determined by certain symmetry principles. Each type of symmetry is associated with a particular conservation law. Space-time symmetry (or Poincaré symmetry) leads to the laws of conservation of momentum and energy. These laws have been verified by direct measurement on every level from the cosmic to the subnuclear. Lorentz symmetry describes the structure of space-time in such a way that the principle of relativity emerges as a consequence. The symmetry principles as a whole tell us what kind of actions may take place and what kind of actions are forbidden. Once verified they are not going to change.

How do we know that these principles are not going to change? Strong evidence for the permanency of experimentally verified physical laws is furnished by telescopic investigation of distant stars. The spectra of light emitted by them gives us information about the kinds of atoms making up these stars. It also gives us information about the interactions between the particles that make up these atoms. What our observations tell us is that all the stars, no matter how far away, are made of the same kind of atoms we have here on earth, interacting by the same forces and according to the same laws of physics that control matter on our planet at the present time. We know that the distant stars are millions and even billions of light-years away. Thus we know, by direct evidence, that the laws which operate here and now are the same laws that existed early in the history of the universe. They have not changed in all the time that has elapsed since very soon after the beginning.

You may argue that this evidence only tells us the laws have not changed in the past. What evidence is there about the future? The evidence is our observation that there is nothing special about our little planet here in this corner of a minor galaxy at this instant of

Discovering the Natural Laws

time. The history of the human race has occupied and will occupy a minuscule period of time compared to the entire lifetime of the universe. So it would be quite remarkable if the laws of nature would change in the near future just because we happen to be here right now. It would be totally arrogant for humans to think that they could have such an effect on the cosmos! (This is not the same as the anthropic principle, which merely argues that we find nature to be constructed the way it is because if it were much different we could not exist to observe it.)

The constancy of the natural laws is fundamental to all physics and to cosmology. Without good reasons that force us to change our minds, the basic laws of physics are not going to be overthrown in the future.

This is not to say that all is now known and that there are no new discoveries to be made. There are still questions to be answered concerning the quantum and the quark. There are questions about the unification of the fundamental forces. There are questions about why the forces and the particles are the way they are and behave the way they behave. But there is no doubt that the fundamental forces are electroweak, gravitational, and strong nuclear. There is no doubt that on the atomic level matter consists of electrons, protons, and neutrons, plus a number of other things. This kind of knowledge will not change. Once we define what we mean by an electron and establish methods for measuring its mass, electric charge, and other properties, the light from distant stars assures us that these properties have not changed in the past. And if they have not changed in the past, why should they change in the future?

Myth 4: *Advanced civilizations of the future will have the use of forces unknown to us at present.*

This has been a favorite tenet of science fiction since its inception. It implies that there are forces in nature still to be discovered in the future and that if our civilization becomes advanced enough we will be able to control these forces and put them to use. These forces will enable the performance of miraculous feats: the elimination of gravity, the moving of planets from one place to another, instantaneous telepathic communication over vast distances, and, most seductive of all, faster-than-light travel to the other stars—even to the other galaxies.

Epilogue (1989)

It is difficult to make a completely rigorous denial of these miraculous forces. The reason I doubt that that these events are going to take place is that there seems to be no room for them in nature. Let me explain what I mean.

We now have identified a very few fundamental forces that control the actions of all matter and energy in the universe. If we divide the electroweak force into its two separate components (the electromagnetic and the weak nuclear force), then there are four of these basic forces. These forces are observed as interactions between pairs of elementary particles. They are characterized by three major properties: their relative strength, the way the strength varies with the distance between the interacting particles (the range), and the property of the particles that determines the nature of the force. For example, the gravitational force depends on the masses of the interacting particles, while the electromagnetic force depends on the electric charge and velocity of the two particles. Both of these are long-range forces, the gravitational force being much weaker than the electromagnetic force.

For the purposes of this discussion let us focus on the relative strength and the range of the various forces. Two particles interacting by means of a long-range force feel an effect whose strength varies inversely as the square of the distance between them. Short-range forces are effective only at very small distances—within the diameter of an atomic nucleus. The following matrix shows how the four forces are classified:

	STRONG	WEAK
LONG RANGE	electromagnetic	gravitational
SHORT RANGE	strong nuclear	weak nuclear

It will be noticed that there is one strong long-range force, one weak long-range force, one strong short-range force, and one weak short-range force. The question we must answer is this: Will a future civilization be able to discover another strong force that is long-range and not already in the above table?

The reason I specify a strong force is that in order for a force to be useful on a macroscopic, human level it must be strong enough to be effective with man-made machinery. A weak force such as the gravitational force requires the mass of an entire planet to produce important effects.

Discovering the Natural Laws

Furthermore, if the force is to move large objects or transmit messages through space, it must be a long-range force. So our question reduces to this: What is the chance of discovering a new and different force that is both strong and long-range?

My answer is simply this: If another strong long-range force existed in nature, we would already have detected it. This would appear to be a rash statement, based on no evidence. However, as we shall see, it is actually embedded within the standard model of fundamental particles, although rarely stated so boldly.

The standard model begins with the observation we have made above: that the four fundamental forces operate as interactions between pairs of particles. These particles make up all matter. The known forces serve to explain the observed behavior of the known particles as they interact with each other. Whenever two electrons approach each other they behave precisely according to the equations describing the electromagnetic force. The same may be said of protons—although when two protons collide with each other at a high-enough velocity the strong nuclear force must also be invoked. Nobody has observed electrons and protons behaving in such mysterious ways that a new and hitherto unknown strong and long-range force must be invoked as explanation.

Fundamental to the standard model is the fact that we never create forces out of nothing. Forces are always the result of interactions between existing particles. All that humans ever do is to use existing forces to rearrange these particles. For example, when we remove a few electrons from a piece of glass by rubbing it with fur, we notice that it has "acquired" a positive electric charge as a result of which we can observe an electric force exerted on another charged object. But that acquired electric charge was not newly created. It was already there within the atoms making up the glass piece. We did not create a new electric force. We simply exposed it to view. The observed electric force is a property of the electric charges already in existence.

What does this example tell us about the creation of hitherto unknown forces? It tells us that in order to create a new strong long-range force we would have to create new types of particles that exist nowhere in the universe. Because all the known forces are completely accounted for by all the currently known particles. There is no room for another force.

Epilogue (1989)

The above argument against the discovery of new strong long-range forces may not be convincing to those who eagerly desire the existence of these forces. To them it is not enough to say that if such a force existed it would already have been observed. However, the burden of proof is on the believers. It is their task to show how it is possible for important new forces to fit into the standard model of elementary particles.

It may also be mentioned that physicists do not deny the possibility of new forces occurring in nature beyond the four listed above. However, these are always assumed to be either short-range forces or, if long-range, then very weak, with effects so small as to be on the very edge of detectability. For example, a major search is currently under way to seek a new component of the gravitational force—either a repulsive force or one that departs from inverse-square behavior or one that depends on the number of baryons (protons and neutrons) in the masses interacting. The most this search has revealed, however, is a force with a strength of perhaps one percent of that of the normal gravitational force, already the weakest force in nature.[7] Indeed, the experimental evidence is so feeble and conflicting that the search for a new gravitational force is beginning to suffer from the same ailments that plague parapsychology research: promising initial results are offset by later negative results following improvement in procedures.

However, the final answer is not in. The new gravity experiments are not to be confused with parapsychology experiments. The work is truly scientific, is based on consistent theories, and is done with great precision. The affair makes a good example of the manner in which physicists, seeking very small effects, can come to disagreement over experimental results.

Considering our great ability to detect very weak forces, the fact that nobody in physics is looking for a new force that is both strong and long-range indicates that physicists are well aware of the truth of the matter: if such a force existed it would have been observed already.

Myth 5: *Advanced civilizations on other planets will possess great forces unavailable to us on earth.*

This myth is regularly referred to when the subject of unidenti-

Discovering the Natural Laws

fied flying objects (UFOs) arises. The film industry has widely disseminated the popular UFO image as that of a large disk hovering high above the ground with no visible form of support. The most obvious question to raise is: *What holds it up?* Magnetic fields won't work; antigravity (as we have seen in Chapter 9) won't work. What, then, holds up the UFO? At this point the true believer habitually responds: "UFOs come from a planet with a civilization millions of years more advanced than ours. Therefore they must have the use of forces we have not yet discovered."

The response to this argument is an extension of that given under Myth 4. Other planets, other galaxies have the same kinds of matter and the same kinds of forces as ours. If our galaxy has only four fundamental forces, all other galaxies have the same four. Once again, this statement is verified by our analysis of the light coming from those distant galaxies. There is no evidence of strange and new forces operating. Even when we observe very mysterious events occurring in distant galaxies—like enormous amounts of energy being emitted by a quasar—physicists have never concluded that these events were caused by any forces other than those presently known. They have never even attempted to explain these phenomena by appealing to unknown forces. Such an explanation would be no explanation at all unless it could be demonstrated how a new force could exist.

APPENDIX I

Wavelengths, Frequencies, and Energy

Suppose we have an atom or nucleus that has an energy E_0 in its normal, ground state. It is raised to an excited state where it has an energy E_1. The energy required to raise the atom to this excited state is $E = E_1 - E_0$. When the atom returns to its ground state—which it soon does—it gives off a photon with the same amount of energy E. The photon energy is related to the frequency of the wave by the Planck relation $E = h\nu$, where ν is the frequency and h is Planck's constant (6.62×10^{-34} joule-second). In any wave, the frequency multiplied by the wavelength equals the wave velocity, so that in a photon of light we have the relationship between frequency and wavelength: $\lambda\nu = c$, where c is the speed of light. Using this equation together with the Planck relation we obtain the relationship between the energy and wavelength of a photon: $E = hc/\lambda$.

If there is a spread $\Delta\nu$ in the frequencies present in a beam of photons from a source, this corresponds to a spread in energy $\Delta E = h\Delta\nu$. From this equation we find that the relative spread or uncertainty in energy is given by the expression: $\Delta E/E = \Delta\nu/\nu$. At the same time, from the equation $\lambda\nu = c$ we find that $\lambda\Delta\nu + \nu\Delta\lambda = 0$, or: $\Delta\nu/\nu = -\Delta\lambda/\lambda$. This means that a given fractional (or percent) spread in energy is connected with the same fractional spread in wavelength and in frequency. As a result, a Doppler-shift measurement of wavelength spread is equivalent to a measurement of energy spread.

Discovering the Natural Laws

Even if we have a large number of identical nuclei, all existing in the same energy level, this does not mean that they will all have precisely the same amount of energy. There will be a small spread of energies among these nuclei. This energy spread is called the width of the energy level.

It is a very fundamental property of nuclei, related to the wave nature of the nucleons, that the width of the energy level is related to the time in which the nucleus exists in that level. When a nucleus is excited to a certain level, it will generally remain in that level for only a finite period of time. If there are a large number of excited nuclei, some of them will decay earlier and some will decay later, but there is a definite average time—the mean lifetime of the nucleus. (This turns out to be the time for about two-thirds of the nuclei to decay to the ground state.)

If the lifetime of a nuclear energy level is very short, then the energy of this level is not very sharply defined—that is, the level has a large width. This means that the gamma ray photons emitted by this level will have a relatively large spread of energies. On the other hand, the longer the mean lifetime of a level, the narrower is the width of the level—which means that the energies of the photons emitted are more precisely defined. In fact, there is a very definite relationship between the lifetime of a level and the energy spread of the level, given by the Heisenberg uncertainty relation:

$$\Delta E \, \Delta t \geqq h$$

where ΔE is the level width, or uncertainty in the energy, Δt is the mean lifetime of the level, and h is Planck's constant. The symbol $\geqq$ means that the product of the energy spread and the lifetime is at least as great as Planck's constant.

Since the ground state of a nucleus exists for essentially an infinite length of time, the ground state has a perfectly defined energy. On the other hand, consider a level with a lifetime of 10^{-8} second. According to the Heisenberg relationship, the level width is $\Delta E \geqq h/\Delta t = 6.6 \times 10^{-26}$ joule =

Wavelengths, Frequencies, and Energy

4.1×10^{-7} eV. This means that if, for example, the level has an energy of 100 keV (100,000 eV), then the gamma rays emitted by that level will have a spread in energy of 4×10^{-7} eV out of 10^5 eV—meaning a precision of four parts out of 10^{12}.

APPENDIX II

Energy and Momentum Relationships for Moving Particles

Consider a particle with rest-mass m, traveling with velocity v. It has a total relativistic mass M and a total energy E given by the equations

$$E = Mc^2 = \frac{mc^2}{\sqrt{1 - \frac{v^2}{c^2}}} \quad (1)$$

while its momentum p is defined as

$$p = Mv \quad (2)$$

Note that the momentum p is a vector quantity, so that its magnitude is found by taking the sum of the squares of the x, y, and z components, according to the Pythagorean Theorem.

$$\begin{aligned} p^2 &= p_x^2 + p_y^2 + p_z^2 \\ &= M^2(v_x^2 + v_y^2 + v_z^2) \\ &= M^2 v^2 \end{aligned}$$

Squaring Equation (1) and transposing terms we have

$$\begin{aligned} (mc^2)^2 &= E^2\left(1 - \frac{v^2}{c^2}\right) = E^2 - E^2\frac{v^2}{c^2} \\ &= E^2 - M^2c^4\left(\frac{v^2}{c^2}\right) \\ &= E^2 - M^2v^2c^2 \\ &= E^2 - (pc)^2 \end{aligned}$$

Energy and Momentum Relationships

Therefore:
$$E^2 = (mc^2)^2 + (pc)^2$$
$$= (mc^2)^2 + (p_x^2 + p_y^2 + p_z^2)c^2 \quad (3)$$

Equation (3) shows us the relationship between the rest-mass-energy, the momentum, and the total energy of the particle.

NUCLEAR Q-VALUES

Let M_1 and M_2 be the total relativistic masses of the two particles entering into a nuclear reaction, while M_3 and M_4 are the masses of the particles emerging from the reaction. E_1, E_2, E_3, and E_4 are the corresponding total energies. By conservation of mass-energy, we write

$$M_1 + M_2 = M_3 + M_4 \quad (4)$$
and
$$E_1 + E_2 = E_3 + E_4 \quad (5)$$

Define Q to be the energy released from the nuclear reaction. It may be written as the difference between the rest-mass energies of the particles before and after the reaction:

$$Q = E_{01} + E_{02} - (E_{03} + E_{04})$$
or
$$E_{01} + E_{02} = E_{03} + E_{04} + Q \quad (6)$$

The total energy of each particle may be written as the sum of its rest energy E_0 and its kinetic energy T:

$$E = E_0 + T$$
or
$$E_0 = E - T \quad (7)$$

Substitute Equation (7) into Equation (6), making use of Equation (5). We then find that

$$Q = T_3 + T_4 - (T_1 + T_2) \quad (8)$$

That is, the Q-value equals the kinetic energy of the outgoing particles minus the kinetic energy of the incoming particles. In addition, we find from Equation (6), using the mass-energy relationship $E_0 = mc^2$,

Discovering the Natural Laws

$$Q = [m_1 + m_2 - (m_3 + m_4)]c^2 \qquad (9)$$

Comparing Equations (8) and (9) we see that Q can be measured in two ways: as the change in kinetic energy of the particles, or as the change in their rest-masses.

ANALYSIS OF THE D-D REACTION

(Method of Taylor and Wheeler)

Consider a deuteron (H^2) at rest. It is part of a deuterium target being bombarded by a beam of energetic deuterons from an accelerator. An exo-energetic reaction results in the production of a proton (H^1) and a triton (H^3):

$$\begin{array}{c} H^2 + H^2 \longrightarrow H^3 + H^1 \\ (1)\ \ (2) \qquad\ \ (3)\ \ (4) \end{array} \qquad (10)$$

(The numbers in parentheses shown beneath the indicated reaction identify the particular particles taking part in the reaction. These are the numbers that will be found in the subscripts of the equations to follow.)

The incident deuteron is assumed to come in along the x-axis, for the sake of simplifying the geometry. Its momentum is thus labeled p_{1x}, and its kinetic energy T_1 is measured by the magnetic analyzer. The target deuteron, at rest, has momentum $p_2 = 0$. The triton is not observed in this experiment; its momentum p_3 may be in any direction. The proton's kinetic energy T_4 is measured as it leaves the reaction along the y-axis. (That is, the instrument only looks at those protons leaving the reaction at right angles to the incident deuterons; there are other protons going in all directions that are not observed.) For the proton, $p_{4x} = p_{4z} = 0$, and the only component of momentum entering into the equations is p_{4y}.

Conservation of energy equates the total energy before the reaction to the total energy after the reaction:

$$E_1 + E_2 = E_3 + E_4 \qquad (11)$$

Energy and Momentum Relationships

Using conservation of momentum we equate the x, y, and z components of momentum before and after the reaction:

$$p_{1x} + 0 = p_{3x} + 0 \tag{12}$$
$$0 + 0 = p_{3y} + p_{4y} \tag{13}$$
$$0 + 0 = p_{3z} + 0 \tag{14}$$

From these equations we see that

$$p_{1x} = p_{3x}$$
$$p_{3y} = -p_{4y}$$
$$p_{3z} = 0$$

We want to calculate the triton's rest-mass m_3, using Equation (3):

$$(m_3 c^2)^2 = E_3^2 - (p_{3x}c)^2 - (p_{3y}c)^2 - (p_{3z}c)^2 \tag{15}$$

It is customary among particle physicists to give masses in equivalent energy units (for example, the mass of a proton is 938 MeV), while the momentum is given in units of E/c. This system has the advantage of eliminating all the c's from the equations to follow. Equation (15) then becomes

$$m_3^2 = E_3^2 - p_{3x}^2 - p_{3y}^2 - p_{3z}^2 \tag{16}$$

We make use of Equation (11) together with the fact that $E_2 = m_2$. (Since particle 2 is at rest, $T_2 = 0$.) Then

$$E_3 = E_1 + m_2 - E_4$$

and, making use of Equations (12) to (14), we find

$$m_3^2 = (E_1 + m_2 - E_4)^2 - p_{3x}^2 - p_{3y}^2$$
$$= (E_1 + m_2 - E_4)^2 - p_{1x}^2 - p_{4y}^2$$

Expanding this equation and rearranging terms, we have

$$m_3^2 = E_1^2 - p_{1x}^2 + E_4^2 - p_{4y}^2 + m_2^2$$
$$+ 2E_1 m_2 - 2E_1 E_4 - 2m_2 E_4$$

We note that $E_1^2 - p_{1x}^2 = m_1^2$ and $E_4^2 - p_{4y}^2 = m_4^2$ and also that $m_1 = m_2$. Finally, after some rearranging we end up with

$$m_3^2 = m_4^2 + 2\,[(m_1 + E_1)\,(m_1 - E_4)] \qquad (17)$$

where $E_1 = m_1 + T_1$ and $E_4 = m_4 + T_4$. Since all the quantities on the right-hand side of the equation are measured—the m's are obtained from mass-spectrometer measurements, and the T's are gotten with the magnetic analyzers—we can calculate m_3, the triton mass, and compare this result with the independent mass value obtained with mass-spectrometer measurements.

The unique feature of this derivation is that in the final expression the T's are small corrections to the m's, so that even though the T's may be measured to one part in one thousand, the E values are good to one part in ten million.

APPENDIX III

The Michelson-Morley and Kennedy-Thorndike Experiments

We want to contrast the predictions of the ether theory with those of the Theory of Relativity. According to relativity, the speed of light with respect to any observer is a constant (c) regardless of the motion of the source or observer. The ether theory, on the other hand, says that the speed of light is c relative to the ether. If the observer is moving with velocity v relative to the ether, then he sees the light traveling with velocity $c_1 = c - v$ if the light is going in the same direction as his own motion, while the light beam sent in the opposite direction has a speed $c_2 = c + v$ relative to the observer.

Consider the time it takes for a pulse of light to leave a source and be reflected from a mirror back to the starting point. In Figure III-1 (a) we show such a light beam, with the apparatus moving parallel to the light path. The time for the round trip is:

$$T_1 = \frac{L_1}{c_1} + \frac{L_1}{c_2}$$

$$= L_1 \left(\frac{1}{c-v} + \frac{1}{c+v} \right)$$

$$= 2L_1 \frac{c}{c^2 - v^2} = \frac{2L_1}{c} \left(\frac{1}{1 - v^2/c^2} \right)$$

The average speed for the light beam over the round trip is seen to be

$$\bar{c} = \frac{2L_1}{T_1} = c \left(1 - \frac{v^2}{c^2} \right)$$

Discovering the Natural Laws

The change in c caused by the motion of the apparatus is seen to be a second order effect, and would not be noticed unless v were very large. If we take v to be 30 km/sec (the earth's orbital velocity), then the difference $c - \bar{c}$ is found to be only 0.003 km/sec, which is too small to be detected by a straightforward measurement of the speed of light.

Now consider another case where the light beam travels at right angles to the motion of the apparatus—as seen by an observer traveling along with that apparatus. Actually, to an

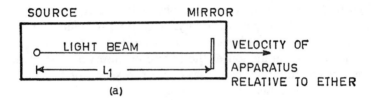

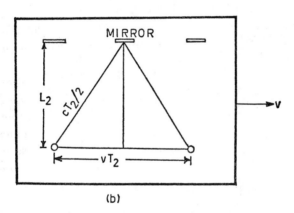

Figure III-1. (a) One-half of a Michelson-Morley apparatus. The light beam travels to and from a mirror while the entire apparatus moves parallel to the path of the light beam. (b) The other half of the apparatus. To the observer moving with the apparatus the light beam is traveling perpendicular to the mirror and perpendicular to the direction of motion. To the observer fixed with respect to the ether, the light beam bounces back and forth along the hypotenuse of a triangle.

Michelson-Morley and Kennedy-Thorndike Experiments

observer at rest with respect to the ether, the light travels a diagonal path, because by the time the light pulse gets back to its starting point, that point has moved horizontally a distance vT_2, where T_2 is the time required for this round trip [Figure III-1 (b)]. The speed of the light pulse along the hypotenuse is c, so that the distance along the hypotenuse is $\frac{1}{2}cT_2$—one-half the total distance traveled in the ether reference frame.

The height of the triangle is L_2, the distance between source and mirror as seen by the observer traveling along with them. By the Pythagorean theorem we can say

$$\left(\frac{cT_2}{2}\right)^2 = L_2{}^2 + \left(\frac{vT_2}{2}\right)^2$$

from which we find

$$T_2 = \frac{2\,L_2/c}{\sqrt{1-(v/c)^2}}$$

We see that the light does not take the same amount of time to travel along the two paths according to the ether theory, since T_1 and T_2 are not equal.

We can think of L_1 and L_2 as the two arms of a Michelson interferometer. The source in that case is the half-silvered mirror from which the two beams diverge and to which they return. The difference in time $\Delta T = T_2 - T_1$ determines whether the two light beams will come together in phase or out of phase at a given point on the photographic film, and we find that

$$\Delta T = T_2 - T_1$$
$$= \frac{2}{c}\left[\frac{L_2}{\sqrt{1-\beta^2}} - \frac{L_1}{1-\beta^2}\right]$$

where we have used the common symbol $\beta = v/c$. Notice that the term describing the "cross-stream" motion of the light has a square-root sign, while the other has not.

Discovering the Natural Laws

Now to look for a change or shift in the interferometer fringes, we turn the apparatus through ninety degrees so that now L_1 is the length of the cross-stream arm. The new round-trip time difference is

$$\Delta T' = T_2' - T_1'$$

$$= \frac{2}{c}\left[\frac{L_2}{1-\beta^2} - \frac{L_1}{\sqrt{1-\beta^2}}\right]$$

The difference between the two ΔT's will tell how much of a fringe shift there will be, for the number of wavelengths along the path of a given beam is

$$N = L/\lambda = L\nu/c = T\nu$$

(where λ = wavelength and ν = frequency), and the difference between the number of wavelengths in the two paths is $\Delta N = \nu \Delta T$. The number of fringes shifted upon the rotation of the interferometer is the second difference $\Delta N' - \Delta N = \nu(\Delta T' - \Delta T)$. We find that

$$\Delta T' - \Delta T = \frac{2}{c}\left[\frac{L_1 + L_2}{1-\beta^2} - \frac{L_1 + L_2}{\sqrt{1-\beta^2}}\right]$$

$$\cong \frac{(L_1 + L_2)}{c}\left(\frac{v^2}{c^2}\right)$$

and $$\Delta N' - \Delta N \cong \frac{L_1 + L_2}{\lambda}\left(\frac{v^2}{c^2}\right)$$

where we have made use of the fact that $\nu = c/\lambda$.

We see from this equation that the ether theory predicts a fringe shift that depends on the ratio of the arm length to the light wavelength, as well as on the ratio v^2/c^2. The M-M experiment showed no measurable fringe shift (or at least a shift much less than you would expect). You can explain this, as Einstein did, by saying outright that the assumptions of the ether theory are false, and that the reason there is no fringe shift is that the speed of light is the same

Michelson-Morley and Kennedy-Thorndike Experiments

along both arms of the interferometer so that $T_1 = T_2$. Or you can take the attitude of Lorentz and FitzGerald and many other people down to this present day, saying that the ether theory is basically correct and that the speed of light (as seen in the moving frame) does vary according to the direction of propagation, but that the length of the interferometer arm changes to compensate for the variation, so that the speed of light simply *appears* to be constant.

Let us see where this theory takes us. Let L_1^0 and L_2^0 be the lengths of the two arms when they are at rest. Now assume that $L_1 = L_1^0 \sqrt{1 - \beta^2}$, and $L_2 = L_2^0$. This tells us that the arm along the direction of motion is shortened, while the other is not affected. Then we find

$$\Delta T = \frac{2}{c} \left[\frac{L_2^0}{\sqrt{1 - \beta^2}} - \frac{L_1^0 \sqrt{1 - \beta^2}}{1 - \beta^2} \right]$$

$$= \frac{2}{c} \frac{(L_2^0 - L_1^0)}{\sqrt{1 - \beta^2}}$$

Upon rotation of the apparatus, L_2 is the shortened arm, and we have

$$\Delta T' = \frac{2}{c} \left[\frac{L_2^0 \sqrt{1 - \beta^2}}{1 - \beta^2} - \frac{L_1^0}{\sqrt{1 - \beta^2}} \right]$$

$$= \frac{2}{c} \frac{(L_2^0 - L_1^0)}{\sqrt{1 - \beta^2}}$$

We then find that $\Delta T' - \Delta T = 0$, regardless of the arm length, and in this manner we have explained away the null result of the M-M experiment.

But now there is another factor to consider: Suppose we wait twelve hours or six months so that the earth has turned around on its axis or the earth has gone halfway around its orbit. If we believe in the ether theory we would expect the speed of the laboratory relative to the ether to be different than before, because the rotary motion of the earth may add or subtract from its orbital motion, or the orbital speed may add or subtract from the sun's galactic speed. We call the new earth velocity v', and $\beta' = v'/c$. Now $\Delta T'$ is not equal to ΔT, and we have a fringe shift (between the first exposure and the second, later exposure) given by the equation

$$\Delta T' - \Delta T = \frac{2}{c}\left[\frac{L_2^0 - L_1^0}{\sqrt{1-\beta'^2}} - \frac{L_2^0 - L_1^0}{\sqrt{1-\beta^2}}\right]$$

$$= \frac{2}{c}(L_2^0 - L_1^0)\left[\frac{1}{\sqrt{1-\beta'^2}} - \frac{1}{\sqrt{1-\beta^2}}\right]$$

We make use of the binomial expansion, neglecting higher-order terms, and find that approximately:

$$\Delta T' - \Delta T \cong \frac{(L_2^0 - L_1^0)}{c}(\beta'^2 - \beta^2)$$

so that the amount of fringe shift will be

$$\Delta N' - \Delta N \cong \frac{\nu}{c}(L_2^0 - L_1^0)\ (\beta'^2 - \beta^2)$$

We see from this that if $L_1^0 = L_2^0$, there will be no fringe shift due to the earth's motion through space. Therefore if we use an interferometer with equal arms we cannot test the hypothesis that the speed of light is independent of the motion of the source—we will always get a null result.

For this reason we use an interferometer with arms of different lengths, as in the Kennedy-Thorndike experiment. And, as was found in that experiment, the result is still no fringe shift. Once again we can say this is what you would expect if the speed of light is simply a constant. But if we want to keep the ether theory we can still patch it up by supposing that the frequency of the atoms emitting the light changes with velocity so that $\nu = \nu_0\sqrt{1-\beta^2}$. (Here ν_0 is the frequency of the clock at rest, while ν is its frequency when moving at velocity v.) We then find that for each position of the interferometer we will have

$$\Delta N = \nu \, \Delta T$$
$$= \nu_0 \sqrt{1 - \beta^2} \cdot \frac{2}{c} \frac{(L_2{}^0 - L_1{}^0)}{\sqrt{1 - \beta^2}}$$
$$= \frac{2\nu_0}{c} (L_2{}^0 - L_1{}^0)$$

This means that ΔN, the phase shift between the two light beams, no longer depends on the velocity of the apparatus, and the fringes will not shift no matter how you turn the device or what time of day it is. In other words, we have explained away the null result of the K-T experiment by assuming a slowing down of the moving atomic clocks, in addition to the contraction in length. Relativity achieves the same results by assuming from the start that the speed of light is constant; it then follows that the clocks will slow down and the lengths will contract.

REFERENCES

Introduction

1. M. A. Rothman, *The Laws of Physics*. New York: Basic Books, (1963).
2. C. N. Yang, *Elementary Particles*. Princeton, New Jersey: Princeton University Press (1962).
3. A. Baker, *Modern Physics and Antiphysics*. Reading, Massachusetts: Addison-Wesley Publishing Co. (1970).
4. K. R. Popper, *The Logic of Scientific Discovery* (2nd edition). New York: Harper & Row (1968).

Chapter 1

1. K. R. Popper, *The Logic of Scientific Discovery* (2nd edition). New York: Harper & Row (1968).
2. Sir Isaac Newton, as quoted by Charles C. Gillispie in *The Edge of Objectivity*. Princeton, New Jersey: Princeton University Press (1960), p. 119.

Chapter 2

1. Mary Hesse, *Am. J. Physics* 32, 905 (1964).
2. Ernst Mach, *The Science of Mechanics*. Chicago: Open Court Publishing Co. (1919).
3. Henri Poincaré, *The Foundations of Science*. Lancaster, Pennsylvania: The Science Press (1946).
4. Hans Reichenbach, *The Philosophy of Space and Time*. New York: Dover Publications (1958).
5. Leigh Page, *Introduction to Theoretical Physics*. New York: D. Van Nostrand Co. (1935).
6. R. B. Lindsay and H. Margenau, *Foundations of Physics*. New York: John Wiley & Sons (1936).
7. Max Jammer, *Concepts of Force*. New York: Harper Torchbooks (1962), and *Concepts of Mass*. New York: Harper & Row (1964).
8. Leonard Eisenbud, *Am. J. Physics* 26, 144 (1958).
9. Robert Weinstock, *Am. J. Physics* 29, 698 (1961).
10. A. B. Arons, *Development of Concepts of Physics*. Reading, Massachusetts: Addison-Wesley Publishing Co. (1965).

11. Sir Isaac Newton, *Principia*, F. Cajori, Ed., Vol. 1, p. 2. Berkeley, California: University of California Press (1966).
12. D. W. Sciama, *The Unity of the Universe*. Garden City, New York: Doubleday & Company (1961).
13. E. F. Taylor and J. A. Wheeler, *Spacetime Physics*. San Francisco: W. H. Freeman and Co. (1966), p. 180.
14. D. W. Sciama, *The Physical Foundations of General Relativity*. Garden City, New York: Doubleday & Company (1969).
15. E. W. Cowan, *Basic Electromagnetism*. New York: Academic Press (1968), p. 5.

Chapter 3

1. Max Jammer, *Concepts of Force*. New York: Harper Torchbooks (1962).
2. E. F. Taylor and J. A. Wheeler, *Spacetime Physics*. San Francisco: W. H. Freeman and Co. (1966), Chapter 3.
3. R. T. Weidner and R. L. Sells, *Elementary Modern Physics*. Boston: Allyn & Bacon (1968), p. 103.

Chapter 4

1. George Gamow, *Gravity*. Garden City, New York: Doubleday & Company (1962).
2. For further information see D. W. Sciama, *The Unity of the Universe*. Garden City, New York: Doubleday & Company (1961). Dr. Sciama has written with great insight on the relationship between gravity and inertia.
3. R. H. Dicke, "The Eötvös Experiment," *Scientific American*, Vol. 205 (1961), p. 84.
4. ———, *The Theoretical Significance of Experimental Relativity*. New York: Gordon & Breach (1964), p. 9.
5. W. M. Kaula, "The Shape of the Earth," in *Introduction to Space Science*, W. N. Hess, ed. New York: Gordon & Breach (1965).
6. D. W. Sciama, *The Physical Foundations of General Relativity*. Garden City, New York: Doubleday & Company (1969).
7. A. S. Eddington, *Space, Time, and Gravitation*. Cambridge: University of Cambridge Press (1921); Chapter VI reprinted in *Readings in the Literature of Science*, W. C. and M. Dampier, eds. New York: Harper Torchbooks (1959).
8. Max Jammer, *Concepts of Force*. New York: Harper Torchbooks (1962), p. 258.

References

9. R. Adler, M. Brazin, and M. Schiffer, *Introduction to General Relativity*. New York: McGraw-Hill Book Company (1965).
10. R. H. Dicke, *The Theoretical Significance of Experimental Relativity, op. cit.*, Appendices 5 and 7.

Chapter 5

1. R. Dugas, "The Birth of a New Science: Mechanics," *The Beginnings of Modern Science*, R. Taton, ed. New York: Basic Books (1964).
2. E. J. Dijksterhuis, *The Mechanization of the World Picture*. New York: Oxford University Press (1964), p. 411.
3. M. A. Rothman, *The Laws of Physics*. New York: Basic Books (1963), Appendix 1.
4. R. Dugas, *op. cit.*, p. 266.
5. E. F. Taylor and J. A. Wheeler, *Spacetime Physics*. San Francisco: W. H. Freeman and Co. (1966), Chapter 2.
6. Sir Isaac Newton, *Principia Mathematica*, revised by F. Cajori. Berkeley, California: University of California Press (1966), p. 23.
7. This relationship is derived in any good physics text; e.g., A. B. Arons, *Development of Concepts in Physics*. Reading, Massachusetts: Addison-Wesley Publishing Co. (1965), p. 392.
8. H. Goldstein, *Classical Mechanics*. Reading, Massachusetts: Addison-Wesley Publishing Co. (1950), p. 4.
9. E. M. Rogers, *Physics for the Inquiring Mind*. Princeton, New Jersey: Princeton University Press (1960), p. 434.
10. Max Jammer, *The Conceptual Development of Quantum Mechanics*. New York: McGraw-Hill Book Company (1966).
11. F. W. von Schelling, *Von der Weltseele*. Hamburg, 1798. Also, *The Encyclopedia of Philosophy*, P. Edwards, ed. New York: The Macmillan Company and The Free Press (1967).
12. E. M. Rogers, *op. cit.*, p. 439.
13. F. Hoyle, *Astronomy*. Garden City, New York: Doubleday & Company (1962), p. 300.
14. Hong-Yee Chiu, *Stellar Physics*. Waltham, Massachusetts: Blaisdell Publishing Co. (1968).
15. P. G. Bergmann, *The Riddle of Gravitation*. New York: Charles Scribner's Sons (1968), p. 176.
16. M. A. Rothman, *op. cit.*, p. 90; H. Goldstein, *op. cit.*, p. 47; also, M. A. Rothman, "Conservation Laws and Symmetry," *Encyclopedia of Physics*, R. M. Besancon, ed. New York: Reinhold Publishing Corp. (1966).

Chapter 6

1. Bela A. Lengyel, *Lasers, Generation of Light by Stimulated Emission.* New York: John Wiley & Sons (1962) and John R. Pierce, *Quantum Electronics,* Garden City, New York: Doubleday & Company (1966).
2. R. H. Kingston, "Laser," *Encyclopedia of Physics,* R. M. Besancon, ed. New York: Reinhold Publishing Corp. (1966).
3. Lengyel, *op. cit.,* p. 99.
4. A complete bibliography of early papers on the Mössbauer effect may be found in *Mössbauer Effect—Selected Reprints,* AAPT Committee on Resource Letters, New York: American Institute of Physics (1963).
5. G. Feinberg and M. Goldhaber, "The Conservation Laws of Physics," *Scientific American* 209, 36 (October 1963).
6. D. E. Nagle, P. P. Craig, and W. E. Keller, "Ultra-high Resolution Gamma-ray Resonance in Zn^{67}," *Nature,* 186, 707 (1960).
7. H. de Waard and G. J. Perlow, *Phys. Rev. Letters* 24, 566 (1970).
8. G. E. Bizina, A. G. Beda, N. A. Burgov, and A. V. Davydov, *Rev. Mod. Phys.,* 36, 358 (1964). Also *Sov. Phys. JETP* 18, 973 (1964).
9. M. C. Hudson and W. H. Johnson, "Atomic Masses of Fe^{56} and Fe^{57}," *Phys. Rev.,* 167, 1064 (1968).
10. E. F. Taylor and J. A. Wheeler, *Spacetime Physics.* San Francisco: W. H. Freeman and Co. (1966), p. 126.
11. D. M. Van Patter and W. W. Buechner, *Phys. Rev.* 87, 51 (1952).
12. S. de Benedetti, C. E. Cowan, W. R. Konneker, and H. Primakoff, *Phys. Rev.* 77, 205 (1950).
13. R. Evans, *The Atomic Nucleus.* New York: McGraw-Hill Book Company (1955), p. 542.
14. C. L. Cowan, Jr., F. Reines, F. B. Harrison, H. W. Kruse, and A. D. McGuire, *Science* 124, 103 (1956); *Phys. Rev.* 117, 159 (1960); also F. Reines, *Science* 141, 778 (1963).

Chapter 7

1. G. Holton, Resource Letter SRT-1 on Special Relativity Theory, *Amer. Journal of Physics,* reprinted as *Special Relativity Theory.* New York: American Institute of Physics (1962).

References

2. R. Resnick, *Introduction to Special Relativity*. New York: John Wiley & Sons (1968).
3. D. I. Blokhintsev, "Basis for Special Relativity Theory Provided by Experiments in High Energy Physics," *Soviet Physics Uspekhi* 9, 405 (1966) (English translation).
4. A. A. Michelson, *Am. J. Sci.* 122, 120 (1881); A. A. Michelson and E. W. Morley, *Am. J. Sci.* 134, 333 (1887).
5. A. Einstein, "On the Electrodynamics of Moving Bodies," *Annalen der Physik* 17, 891 (1905), translated and reprinted in *The Principle of Relativity*. New York: Dover Publications.
6. H. P. Robertson, "Postulate versus Observation in the Special Theory of Relativity," *Rev. Mod. Phys.* 21, 378 (1949).
7. R. J. Kennedy and E. M. Thorndike, "Experimental Establishment of the Relativity of Time," *Phys. Rev.* 46, 400 (1932).
8. H. E. Ives and G. R. Stillwell, *Journal of the Optical Society of America* 28, 215 (1938); H. E. Ives and G. R. Stillwell, *J. Opt. Soc. Am.* 31, 369 (1941).
9. T. S. Jaseja, A. Javin, and C. H. Townes, *Phys. Rev.* 133, A1221 (1964).
10. J. P. Cedarholm, G. F. Bland, W. W. Havens, Jr., and C. H. Townes, *Phys. Rev. Letters* 1, 342 (1958); also J. P. Cedarholm and C. H. Townes, *Nature* 184, 1350 (1959).
11. H. J. Hay, J. P. Schiffer, T. E. Cranshaw, and P. A. Egelstaff, *Phys. Rev. Letters* 4, 165 (1960).
12. D. C. Champeney, G. R. Isaak, and A. M. Khan, *Physics Letters* 7, 241 (1963).
13. A survey of atomic physics experiments having a bearing on the fundamental laws of physics, including those of relativity, is given by G. Feinberg in *Atomic Physics*, V. W. Hughes, B. Bederson, V. W. Cohen, and F. M. J. Pichanick, eds. New York: Plenum Publishing Corporation (1969).

Chapter 8

1. Rosa and Dorsey, *Bur. Standards Bull.* 3, 433 (1907). Also described in Harnwell and Livingood, *Experimental Atomic Physics*. New York: McGraw-Hill Book Company (1933).
2. E. M. Purcell, *Electricity and Magnetism*. New York: McGraw-Hill Book Company (1965).
3. R. A. Lyttleton and H. Bondi, *Proc. Roy. Soc.* (London) A252, 313 (1959).

4. R. W. Stover, T. I. Moran, and J. W. Trischka, *Phys. Rev.* 164, 1599 (1967).
5. J. G. King, *Phys. Rev. Letters* 5, 562 (1960).
6. A. M. Hillas and T. E. Cranshaw, *Nature* 184, 892 (1959).
7. S. J. Plimpton and W. E. Lawton, *Phys. Rev.* 50, 1066 (1936).
8. A. S. Goldhaber and M. M. Nieto, "A New Geomagnetic Limit on the Mass of the Photon," *Phys. Rev. Letters* 21, 567 (1968).

Chapter 10

1. *The Skeptical Inquirer* is the journal of The Committee for the Scientific Investigation of Claims of the Paranormal.
2. Milton A. Rothman, *A Physicist's Guide to Skepticism*. Buffalo, New York: Prometheus Books (1988).
3. Bertrand Russell, *Human Knowledge, Its Scope and Limits*. New York: Simon & Schuster (1948), p. 381.
4. Jacob Bronowski, *The Ascent of Man*. Boston: Little, Brown and Company (1973), Chapter 11.
5. Rothman, *op. cit.*, Chapter 6.
6. Thomas S. Kuhn, *The Structure of Scientific Revolutions*. Chicago: The University of Chicago Press (1962).
7. Bertram Schwartzschild, "From Mine Shafts to Cliffs—the 'Fifth Force' Remains Elusive," *Physics Today* 41, no. 7, 21 (July 1988).

Index

(Separate Index to Chapter 10 follows.)

Acceleration
  Einstein and, 79
  in orbiting body, 45
Annihilation of particles, 124–30
Arons, A. B., 18
Atoms
  excited states of, 101–6
  hydrogen, structure of, 49–50

Bainbridge, K. T., 121
Berkeley, Bishop George, 25
Bio-Savart Law, 161, 168
Bizina, G. E., 115
Bland, G. F., 154
Blokhintsev, D. I., 131
Bondi, Hermann, 169
Brahe, Tycho, 6–7
Brans, C., 81–82
Brans-Dicke theory, 81–82

Cavendish, Henry, 62, 174
Cedarholm, J. P., 154
Centrifugal force, 23
Champeney, D. C., 157
Collisions, momentum and, 83–91
Conservation laws, 60
  of electric charge, 168–72
  historical background to, 83–100
  modern experiments on, 101–30
Conservation of Energy, Law of, 49–52, 96–98, 100, 114
Conservation of Mass, Law of, 119–20
Conservation of Momentum, Law of, 29–30, 33-34, 85
Conservation of Motion, Law of, 88
Coriolis force, 23
Coulomb, Charles Augustin, 47, 173–74
Coulomb's Law, 79, 160, 164, 167, 173–82
Cranshaw, T. E., 155, 171
Creation of particles, 124–30

D-D Reaction, analysis of, 206–8
Deduction, process of, defined, 9
Descartes, René, 84–86, 88
Deuterons, formation of, 56
Dicke, Robert H., 66–67, 70, 81–82
Dirac, Paul, 124
Doppler broadening of the spectrum line (Doppler effect), 104–5, 114, 135, 149, 151, 155–57

Dorsey, 163

Eddington, A. S., 79
Egelstaff, P. A., 155
Einstein, Albert, 8, 25, 37, 44, 79–82, 123
  field equations of, 80
  General Theory of Relativity, 25, 46, 65, 81, 89, 116
  Principle of Relativity, 131–59, 192
  Special Theory of Relativity, 37, 131, 136–38
Eisenbud, Leonard, 16, 20
Electric charges
  equality and conservation of, 168–72
  invariance of, 172–73
  quantized, 61
Electromagnetic energy, 52
Electromagnetic force
  adequate knowledge of, 59
  features of, 53
  law of quantum mechanics and, 60
  particle response to, 61
  photons and, 52, 54
  relative strength of, 58
Electromagnetism, 160–82
Electrons, characteristics of, 12
Electrostatic force, 53–54
  Coulomb's Law of, 79, 160, 164, 167, 173–82
Energy
  conservation of, ix–x, 1–2, 92–99
  electromagnetic, 52
  frequencies, wavelengths and, 201–3
  heat, 51
  hypotheses on, 10–11
  momentum relationships for moving particles and, 204–8
Energy, Law of Conservation of, 49–52, 96–98, 100, 114
Eötvös, Baron Roland von, 65–67, 69
Epistemology, defined, 2
Equality of electric charges, 168–72
Equivalence, Principle of, 65
ESP (extrasensory perception), ix, 193–99
Euclidean geometry, 16, 187
Excited states
  of atoms, 101–6
  of nuclei, 106–16
"Experimental Establishment of the Relativity of Time" (Thorndike and Kennedy), 146
Extrasensory perception (ESP), ix, 193–99

Falling objects, gravitational force and, 63–71
Fermi, Enrico, 129
Fictitious force, 23
Field equations of Einstein, 80
First Law of Motion, 26–27, 83
FitzGerald, George F., 143
"Fixed in space," defined, 73
Forces
  acceleration and, 45
  basic, 52 (see also Electromagnetic force; Electrostatic force; Gravitational force)
  defined, 13–21
  as important concept, 48–49
  inertial, 23
  Laws of Motion and, 22 (see also Motion, Laws of)
  strong and weak nuclear, 52, 54–55, 57, 58
Frequencies, energy, wavelengths and, 201–3

g-force, 23

Index

Galileo, ix, 48, 63–64
Gauss, Johann, 174–75
General Theory of Relativity, 25, 46, 65, 81, 89, 116
Goldhaber, A. S., 178
Gravitation
 Laws of Motion and, 44–47
  (see also Motion, Laws of)
 Universal Law of, 5–8, 62, 64, 71, 76–81
Gravitational field, 79
Gravitational force, 62–82
 defined, 52–53
 falling objects and, 63–71
 inertial mass and, 62–63
 inverse-square law of gravitation and, 71–82
 relative strength of, 57–58
Gravitational mass, 63
Ground state of atoms, 102
Guess and test method, 1–12
 conceptual structure of science and, 8–12
 examples of, 5–8

Havens, W. W., 154–55
Hay, H. J., 156
Heat energy, 51
Heisenberg, Werner K., 114
Hertz, Heinrich, 160, 163
Hesse, Mary, 16
Hillas, A. M., 172
Hoyle, Fred, 97–98
Huygens, Christian, 85–89
Hydrogen atoms, structure of, 49–50
Hypothesis, defined, 4

Inertial forces, 23
Inertial mass, 62–63
Inertial reference frames, 17, 23
 defined, 26–27
Invariance of electric charge, 172–73
Inverse-square law of gravitation, gravitational force and, 71–82
Isaak, G. R., 157
Ives, Herbert E., 148, 151, 152
Ives-Stillwell experiment, 148–52, 155

Jammer, Max, 16, 95
Jaseja, T. S., 153
Javin, A., 153
Joule, James Prescott, ix, 96

Kennedy, Roy J., 144
Kennedy-Thorndike experiment (K-T experiment), 144–47, 152–54, 209–15
Kepler, Johannes, 6–7, 45, 77
Khan, A. M., 157
King, J. G., 171, 173

Lagrange, Joseph-Louis, 92
Laplace, Pierre Simon Marquis de, 95
Laser (Light Amplification by Stimulated Emission of Radiation), 105–6
Lavoisier, Antoine Laurent, 95
Lawton, W. E., 176
Lee, T. D., xi
Legendre Polynomials, 76
Leibniz, Gottfried Wilhelm, 88
Leverrier, Urbain, 78
Life, physics and, 192–99
Light, measuring differences in speed of, 134–36
Lindsay, R. B., 16
Lorentz, Hendrik A., 139, 143
Lorentz transformation equations, 139–40
Lorentz-FitzGerald contraction, 143–44, 147, 152, 159, 166
Lyttleton, R. A., 169

Mach, Ernst, 16, 17–18, 25, 27
Mach's Principle, 25

Margenau, H., 16
Mass
  defined, 13–20
  inertial, 62–63
  Law of Conservation of, 119–20
Maxwell, James Clerk, 162, 174–75
Maxwell's equations, 136, 162–64, 180
Maxwell's Theory of Electromagnetic Radiation, 163
Mercury (planet), 78–82
Michelson, A. A., 134, 163
Michelson-Morley experiment (M-M-type experiment), 136, 138, 140–44, 152, 153, 158, 209–15
Millikan oil drop experiment, 168–69
Momentum
  collisions and, 83–91
  Law of Conservation of, 29–30, 33–34, 85
Moran, T. I., 169
Mössbauer, Rudolf L., 112–13
Mössbauer Effect, 106–16, 155, 156
Motion, Laws of, 13–47
  experimental tests of, 35–44
  First, 26–27, 83
  gravitation and, 44–47
  recreating, 20–33
  Second, 31–32, 39, 46, 48, 64, 72, 83
  Third, 33–35, 43–44, 83

Natural laws, 8, 16
  events allowed and forbidden by, 183–88
Neutrino, 129–30
Newton, Sir Isaac, ix, 9, 65, 131, 133
  Conservation of Momentum and, 89–91

Universal Law of Gravitation, 5–8, 62, 64, 71, 76–81
  See also Motion, Laws of
Nieto, M. M., 178
Nuclear mass differences, 120–21
Nuclear reactions, 116–23
Nuclei, excited states of, 106–16

Orbiting bodies, acceleration in, 45

Page, Leigh, 16
Particle response to electromagnetic force, 61
Particles
  creation and annihilation of, 124–30
  momentum relationships for moving, 204–8
Pauli, Wolfgang, 129
Perigee, defined, 74
Perihelion
  defined, 74
  precession of, 75
Philosophy, scientific knowledge and, 1–2
Photons, electromagnetic force and, 52, 54
Physics
  life and, 192–99
  science fiction and, 188–92
Planetary Motion, Third Law of, 6, 45, 77
Plimpton, S. J., 176
Poincaré, Henri, 16
Popper, Karl, xi, 4, 9
Potential energy well, 106–7
Principia Mathematica (Newton), 5
Principle of Relativity, 131–59, 192
Psychic powers (vital force), 192–93

Index

Q-values, 120–23, 205–6
Quantized electric charges, 61
Quantum electrodynamics, theory of, 54, 59
Quantum mechanics, law of, 60

Reichenbach, Hans, 16
Relativity
  General Theory of, 25, 46, 65, 81, 89, 116
  Principle of, 131–59, 192
  Special Theory of, 37, 131, 136–38
Resnick, R., 131
Resource Letter of Philosophical Foundations of Classical Mathematics (Hesse), 16
Robertson, H. P., 138, 152
Roemer, Ole, 134–35
Rogers, Eric, 96
Rosa, 163

Schelling, Friedrich Wilhelm von, 95–96
Schiffer, J. P., 155
Schroedinger, Erwin, 178
Schwartzschild solution of Einstein's field equations, 80
Sciama, D. W., 37
Science
  as conceptual structure, 8–12
  defined, xii
  philosophy and, 1–2
Science fiction, physics and, 188–92
Scientific knowledge, basis of, 1–4
Second Law of Motion, 31–32, 39, 46, 48, 64, 72, 83
Spacetime Physics (Taylor and Wheeler), 159
Special Theory of Relativity, 37, 131, 136–38

Stevin, Simon, 64
Stillwell, G. R., 148, 151
Stover, R. W., 169
Strong nuclear force, 52, 54–55, 58
Symmetry, xi, 99–100

Tachyons, 191–92
Taylor, E. F., 122–23, 159, 206–8
Third Law of Motion, 33–35, 43–44, 83
Third Law of Planetary Motion, 6, 45, 77
Thompson, Benjamin (Count Rumford), 95, 96
Thorndike, Edward M., 144
Total mechanical energy, 93
Townes, C. H., 153–55
Trischka, J. W., 169

Universal Law of Gravitation, 5–8, 62, 64, 71, 76–81
Universe, view of, 199–200

Van de Graaff generators, 118
Vital force (psychic powers), 192–93

Wallis, John, 85, 86
Waterston, J. J., 95
Wavelengths, frequencies, energy and, 201–3
Weak nuclear force, 52, 57, 58
Weinstock, Robert, 16
Wheeler, J. A., 122–23, 159, 206–8
Wren, Sir Christopher, 85

Yang, C. N., xi
Yukawa, Hideki, 179

Index to Chapter 10

Anthropic principle, 210
Antigravity, 202, 214
Bronowski, Jacob, 203
Conservation laws, 209
Creation science, 202, 206
Descartes, 203
Electroweak force, 211
Energy, conservation of, 201, 207
Faster-than-light travel, 202, 208
Force, psychic, 206
Grand unified field theory, 201
Gravitational force, 211
Heisenberg uncertainty principle, 204
Impossibility, 207, 208
Interactions, 205, 207, 212
Knowledge, perfect, 204
Kuhn, Thomas, 208
Laws of denial, 202
Laws of permission, 202
Long-range force, 211
Matter, standard model of, 201
Mysticism, 201, 207
Myths about science, 203
New age, 202
Newton, Isaac, 205
Paradigm shift, 208
Paranormal phenomena, 202
Parapsychology, 202
Pseudoscience, 202, 203
Science fiction, 202
Short-range force, 211
Skepticism, 203
Standard Model, 205
Strong nuclear force, 211
Superstring theories, 201
Symmetries, 201, 209
Telepathy, 208
UFOs, 202, 214
Weak nuclear force, 211